DG647328

1

10^{25} meters
1 billion light-years

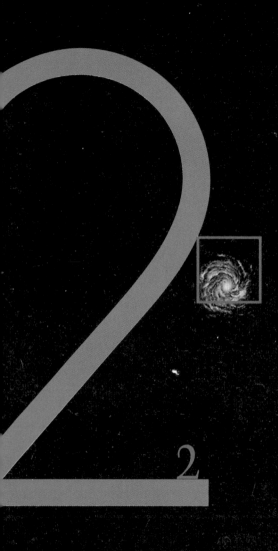

10²² meters
1 million light-years

10⁸ meters
100,000 kilometers

10⁵ meters
100 kilometers

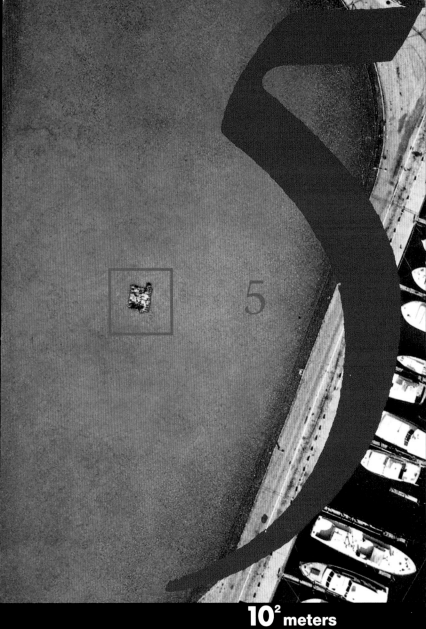

5

10² meters
100 meters

6

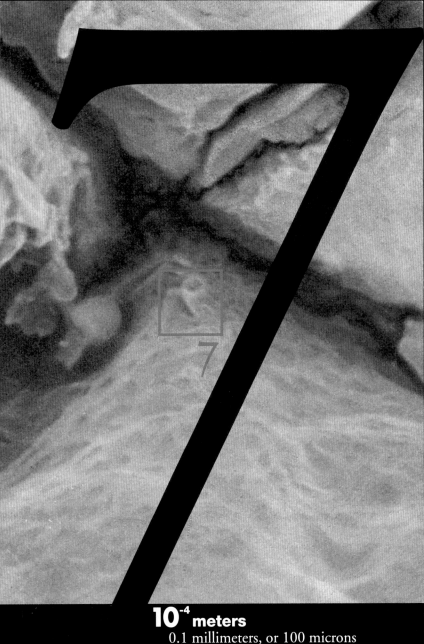

10⁻⁴ **meters**
0.1 millimeters, or 100 microns

10⁻⁷ meters
0.1 micron, or 1,000 angstroms

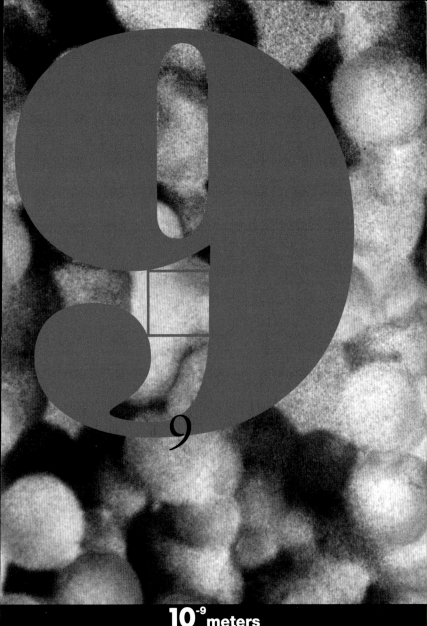

9

10⁻⁹ meters
10 angstroms, or 1 nanometer

CONTENTS

NUMBERS
THE UNIVERSAL LANGUAGE
Denis Guedj

We take numbers for granted and use them every day. Numbers are universal: all cultures have them. Yet the idea of number, so obvious to us today, is the result of a protracted process of progressively abstract reasoning. It took humanity thousands of years to advance from literal quantities to the concept of numbers. How do we add things up? By seeing each object simply and exclusively as a unit. By seeing things as entities while rejecting their specific differences.

CHAPTER 1

HOW MANY?

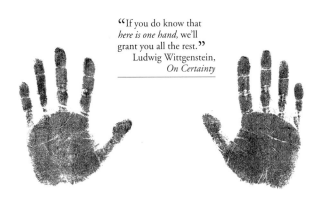

"If you do know that *here is one hand,* we'll grant you all the rest."
Ludwig Wittgenstein,
On Certainty

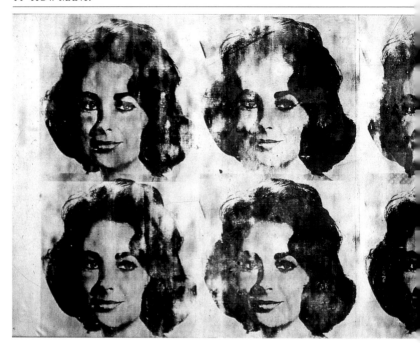

The idea of number

From the point of view of the number—a completely original perspective—all objects are the same but not identical. The idea of number is based on a division of the world into two levels: *the same* and *the different*. The things we enumerate are the same essentially but not numerically—that is, they have the same value but are not one-and-the-same. If they were not different, there would be only one object in the world.

Take a pile of diverse objects mixed together. To note their equal existence and nothing more is to say *how many* they are. In order to do so, it is necessary to state that they exist and that they exist in the same way, none existing differently from the others. At the same time, this affirms that they are discrete from one another, without detailing their differences. Once this principle is established, it is possible to count the objects.

In 1963 the artist Andy Warhol painted *Liz Taylor 10 Times:* ten faces, virtually identical, of the actress Elizabeth Taylor. The reproduction of things that are the same is a human obsession—witness our fascination with the possibilities of cloning. When this theme is treated in art, the inspiration may come not from the repetition itself but from the calculated variations from the identical.

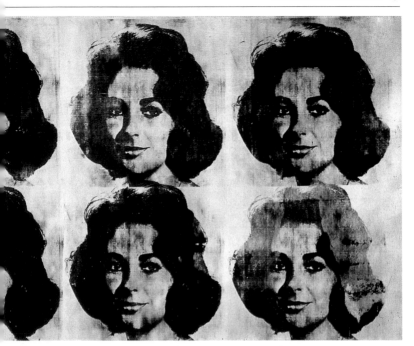

To count a herd of deer, you must suppress any desire to distinguish among them—buck or doe, fawn or yearling or adult—and at the same time hold the conviction that each one of them is separate from the others, that the herd is more than one deer.

Where the eye is weak, the finger is strong

The eyes can recognize many things, perceiving and remembering a host of facial characteristics, for example, or a multitude of landscape features. Yet when it comes to numbers, they exhibit a notable weakness.

Each of us has experienced the problem of registering more than five objects with a single glance. If you were to cut off your fingers and pile them on the table, you would not be able to see, at a glance, if you had lost one in the process.

Because we are incapable of grasping quantity directly by sight, we have invented numbers. And with them,

In mathematics, the set $\{a, a, a, a, a, a, a, a, a, a\}$ contains not ten elements but only one. The only object in the universe that belongs to it, in effect, is a. Thus, in *set theory*, $\{a, a, a, a, a, a, a, a, a\} = \{a\}$. A set with one unit is called a *singleton*. A set is a singleton if $x = y$ no matter what the x or y belonging to it are.

we invented counting. To keep track of quantity, we made marks; then we named the marks and memorized the names.

How do we keep track of quantity?

The most ancient numerical marks occur in the first human civilizations, those of the Paleolithic era. Humans must have learned how to keep track of numbers with the same early intelligence that brought them to preserve and use fire, and presumably at around the same time.

Take a group of things—animals, people, or objects. How can we remember how many there are without any idea of their number? By making a mark, often a notch, for each thing on a sort of document. For each thing, make one mark: this is apparently one of the oldest practices in the world. Wherever they lived, the Paleolithic peoples had a good surface available to them for their documents: bone. In certain areas they also used wood, although bone resists the effects of weather and damp better. Bones incised with numerical notches almost thirty thousand years old have been found.

It is striking that in many early cultures the idea of *many* preceded the idea of *one*. A significant proportion of humankind reversed what we now think of as the traditional relationship in numbers. Thus, in religious history, for example, polytheism often antedated monotheism. For the most part, the gods were multiple before a one god appeared.

Mapping numbers on the body

Besides bone, wood, and stone recording devices, people used their own bodies as a tool to remember quantity. Of course they did not place marks on the body's surface,

B elow: a notched reindeer antler dating from the Upper Paleolithic era (about 15,000 BC). The use of notches began well before recorded history. The tally was a common way of representing credit; seller and buyer each kept an identical board. When the latter took merchandise on credit, the seller stacked the boards one on top of the other and engraved continuous lines representing the amount of merchandise exchanged. The system made cheating impossible: the buyer could not efface a notch, nor could the seller add one.

O pposite: in the 8th century the English monk Bede wrote a treatise on digital calculation in which the fingers of the left hand represented ones and tens, those of the right hand stood for hundreds and thousands, and various positions of the hand in relation to specific parts of the body signified tens and hundreds of thousands.

much less score notches in it. Instead, each body part and position was assigned a number. Not only were fingers used for counting, but also ears, arms and legs, the torso, the head, individual finger joints, and knuckles. Many civilizations thus developed highly complex corporal numerical maps, accompanied by a grammar of gesture expressed principally by the

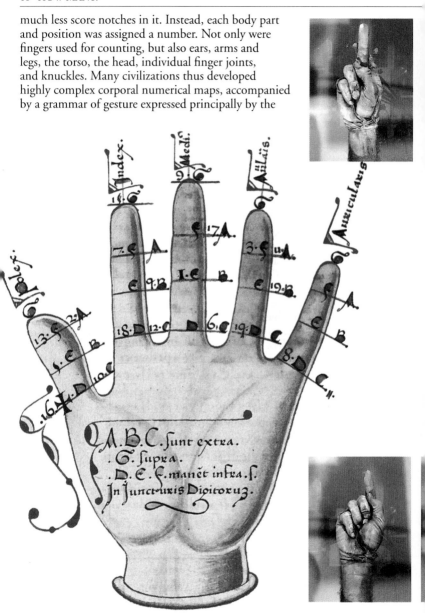

fingers placed in different positions—extended, bent, or curved.

This dancelike use of the fingers is called *digital calculation.* It is an elegantly simple technique with which one can reach unexpectedly impressive figures. In 16th-century China accountants developed a system employing two hands that allowed them to count beyond a billion (that is, 10^9).

This counting method aids the memory, making it possible to keep in mind the particular place on the body where a number of objects belongs, as the science historian Georges Ifrah notes in *From One to Zero: A Universal History of Numbers.* But how did people missing legs, arms, or hands count? When thieves were punished by having a hand cut off, they were forever debarred from counting up their booty.

Natural sequence

While number is intimately connected to objects, whose quantity it represents, the idea of *natural sequence* is

Manuals on digital calculation were published as late as the 15th century (opposite). Today some African peoples continue to use digital calculation. Above and below: a Masai counter.

purely conceptual, having no relation to the concrete natural world. While *two* is related to an eagle's pair of wings, *four* to the legs of an antelope, and *one* to a human being's nose, *two, four,* and *one*—the numbers themselves—have no connection with one another. Since they do not, why place them in a certain order? Why place two before four? The abstract idea of number itself must exist, divorced from specific quantities of things, before we can conceive of the sequence of numbers.

What do we know about this sequence? We can name several of its features:

- numbers follow one another, marching in single file
- another number always tags on after the last
- if we know a number, we know the one that follows it, which is always obtained by adding one
- numbers get bigger and bigger
- they follow one another endlessly
- there is no last number, although there is a first
- they are in order
- they create, in fact, the very archetype of order.

To put a series of objects in order is to assign to them a standard order: first, second, third, and so on.

The source for the idea of sequence may be the ancient act of ticking off the fingers of the hand, because it depends on a morphology that implies a natural

A pair of wings is flying through the sky. A pair of scissors approaches and cuts them in two. The suddenly heavy wings fall to the ground but, soaring in the sky of concepts, the idea of twoness remains. A thought process similar to this, progressing from a number of things to simply the idea of the number itself, gave rise to abstract numbers.

order. It should be noted that the human hand is not arranged with radial symmetry like the spokes of a bicycle wheel. Imagine a circular hand, with fingers extending symmetrically from a palm joined to the wrist at the center of the back of the hand. In such a hand no single digit would impress us as being first. We would have to pick one. Who would do so?

I magine a radial hand connected to the wrist from the center of its back. The five fingers are identical in shape and length, arranged symmetrically around the palm. Now count them! Which one do you start with? If the human hand were designed this way, would we have arrived at the ordinal-cardinal relationship that constitutes the main richness of number?

Number has two faces

In order for the Paleolithic bison hunter to record the killing of four beasts, he or she had to count—that is, to inspect them: first, second, third, fourth. Because they were ticked off to the fourth finger and no further, there were four. This counting action gives us both quantity and sequence together: the two roles are inseparable. When quantity is assessed, the numbers are called *cardinal;* when the sequence is assessed, the numbers are *ordinal.* An ordinal number is seen as the link of a chain; a cardinal is seen as pure quantity. The cardinal measures; the ordinal orders.

Do animals count?

Has number made an impression in the minds of animals? What are the numerical abilities of the various species? Are they capable of recognizing and remembering quantities, perhaps in a rudimentary way? Can they take into account quantitative differences when they think? Indeed, do animals share with humans the concept of quantity, or are humans alone in displaying a numerical

At the beginning of the century it was claimed that Hans the horse (opposite below) could count and even add fractions by tapping his hoof. Several years later it emerged that the horse based its responses on its trainer's signs. This inaugurated a period in which scientists studied the numerical abilities of animals. Left: a researcher works with a chimpanzee. Chimps are adept and intelligent learners and can grasp some counting concepts.

aptitude? Since, as we have seen, number is the product of abstract thinking, the answers to these questions have profound significance.

Some species seem to be capable of developing a certain numeric sense. The female wasp *Genus eumenus* manages to distinguish *five* from *ten* as successfully as she distinguishes male from female eggs. Having placed her eggs in cells, she puts precisely ten caterpillars in each of those containing a female egg, and she puts precisely five in those housing a male egg. Some chimpanzees can indicate the central element in an odd set of objects presented in a row. In one case study a jackdaw, confronted with a row of bowls and receiving a set of

four signals, has indicated the bowl placed fourth in the row, and so on. A great spotted woodpecker was taught a code of request involving numerical order: 1 tap for a pistachio, 2 for a cricket, 3 for a mealworm, 2 + 2 for a mayfly, 2 + 2 + 3 for a locust. The woodpecker responded correctly.

What does this all prove? That animals know how to count? No. In order to do that, they would have to be able to enumerate any kind of series. Up to now, no one has produced a single specimen capable of such a performance. Will humans remain the only species that can proclaim: I count, therefore I am?

"Many birds, for instance, have…a number sense. If a nest contains four eggs one can safely be taken, but when two are removed the bird generally deserts. In some unaccountable way the bird can distinguish two from three."

Tobias Dantzig,
Number: The Language of Science, 1930

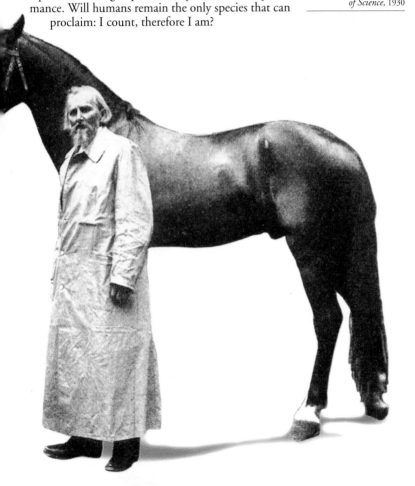

Wat system of representing numbers has the same power and flexibility as the numbers themselves, is capable of following them as far as they can go, and, as soon as the need for a new number arises, can offer it a name? What system is so adept at naming the new?

CHAPTER 2

FROM NUMBERS TO FIGURES

From the stones of monuments to the fabric of clothing, every possible surface has been used to record numbers. On the ancient Egyptian wall at left, on the decorated lottery player at right: are these figures or numbers? Figures, numbers—what's the difference? Do we say a digit has three numbers or a number has three digits?

Constructing numeration

In creating a home for the idea of quantity, certain peoples made do with summary structures in which they housed a fistful of numbers sufficient for their needs: *one, two, three, many.* Others, anxious to welcome the numerical multitude, fashioned elaborate monuments called *numeration systems,* or *numerations.*

A numeration is a system of representation of numbers. The universe of numbers has a characteristic that makes it unique in the realm of human endeavor: it uses a triple system of representation—visual, oral, and written, which respectively employ figural (or symbolic), spoken, and written numerations. See the number, speak the number, write the number: these are the mighty tasks of numerations.

The first function of numerations was to represent individual numbers. They then took on a second function: calculation. It is possible to represent numbers without calculating, but impossible to calculate, even mentally, without representing numbers. Different kinds of numerations serve different purposes, some more sophisticated and complex than others.

Figural numerations are concrete notations, composed of a system of physical marks that appear on solid surfaces. *Spoken numerations* are words; they give a name to each number. When they are transcribed into writing,

The abacus is still used in some parts of the world—above, in Afghanistan.

these words appear entirely in letters: one, two, one hundred, one thousand, and so on. *Written numerations* utilize symbols to represent numbers: 1, 2, 100, 1,000, and so on. The most rudimentary numeration systems do not go beyond simple representation; spoken numerations, for example, have no calculating capacity—you cannot calculate with the word *ten*. What are these various numeration systems, and how do they work?

Figural numerations

Until the invention in India in the 5th century AD of *positional notation,* which

To use an abacus requires both conceptual mastery and manual dexterity, calling into play a complex set of gestures, aesthetic as well as efficient. Sound also contributes to this performance—the click and rattle of the counters striking the wooden frame of the abacus—so that the effect is almost that of a dance. The invention of written calculation put an end to the body's participation in the art of calculating.

included the invention of the zero, the job of calculating was exclusively performed by figural numerations. In this system, each number is represented by a physical symbol. This can be a set of marks on a surface, such as the notches in a Paleolithic tally bone, or a collection of objects—pebbles, pearls, shells, sticks, knots in a string, counters.

The simplest numerations are limited to static position systems; the most complex involve subtle changes of the

This detail from an ancient Greek vase painting shows a payment of tribute to the Persian emperor Darius being recorded.

position of objects. A wide variety of devices was developed to manage these: the computing table, a dust-covered board, the abacus, and other sorts of calculation aids.

Knotted cords, in use in Persia during the reign of Darius in the 5th century BC, served this purpose. Tallying and calculation were done by a technique of knots arranged along cords. By the 13th century the Inca had improved this method with the invention of the *quipu,* or "numbers on strings." A cord was divided by marks indicating various levels of quantity: single units, tens, hundreds. At each

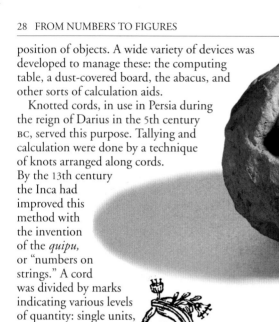

L eft: an Inca accountant uses a record-keeping device called a *quipu,* a cord held horizontally from which knotted strings hang vertically. The information it retains is encoded in the type of knots used, the length of the cord, and the color and position of the strings.

"Soon after, [Darius] called a meeting of the Ionian commanders and showed them a long leather strap in which he had tied sixty knots. 'Men of Ionia…,' he said, 'I want you to take this strap, and every day undo one of the knots.… Should I fail to return before all the knots have given out, you are at liberty to sail home.'"
Herodotus, *The Histories,* Book 4, 5th century BC

level were placed as many knots as needed to record a number. This method, which allows for complex concrete calculation, relies on the concept of positional notation.

The pebble, cornerstone of numerical structure

The primitive practice of pairing is a form of figural numeration. One begins with the simple idea that a single pebble is worth the quantity *one.* To calculate in

this way means dealing with enormous piles, impossible to manipulate. Thus a single pebble of different form, size or perhaps color, came to replace a pile (e.g., a pile of ten or of one hundred pebbles). This called for establishing conventions detailing the value of each type of pebble and setting up a hierarchy among the different types. This may be the origin of the *principle of the base,* the foundation of all numerations.

Instead of natural pebbles, scarce in some countries (for example, Mesopotamia), man-made objects, often of clay, were preferred. Sumerian clay stones known as *calculi* (*calculus* is Latin for stone) have been found that date to the 4th millennium BC. The different shapes of these *calculi*—small and large spheres and cones, perforated or smooth—were the source of the forms given to the written figures of Sumerian numeration.

Calculations with no memory

All these apparatuses suffered from a major flaw: they could not keep a record of the past. Each stage of a calculation in effect wiped out those preceding it. If an error crept in, how could its origin be traced? How could

Calculi are small clay objects of various shapes that represent specific digits in Sumerian numeration, which has a base of 60. The small cone has a value of 1, the ball 10, the large cone 60, the perforated large cone 3,600, and the perforated sphere 36,000. The much larger perforated sphere seal-container, on the page opposite, had a social or contractual rather than an accounting role. When a contract specifying a particular number was agreed upon, the *calculi* representing the sum of that number were placed inside the hollow ball. To make it unnecessary to break the ball in order to learn its contents, notches representing the *calculi* inside were made on the surface of the seal.

the calculation be checked? The entire reckoning would have to start over again. Because these techniques were irreversible they could only be used for temporary operations; without memory, they were fixed on the moment.

Thus, figural numerations offered, on the one hand, marks or objects frozen in time and, on the other, a working calculator that produced results but left no trace of its operations: nonfunctional inscriptions for one, ephemeral operations for the other.

An improvement was the inscription of successive stages of a calculation on a surface. This allowed its rereading at any point and provided the sought-after permanence. The importance to calculation of the invention of paper by the Chinese in the 2d century AD cannot be overestimated.

Spoken numerations

A spoken numeration is a system of the nomination of numbers. *One thousand and one* and *mille et un* are expressions pertaining to spoken numeration, the first belonging to the English language, the second to the French, while *1,001* is a word in written numeration, using positional notation, which is not tied to any particular language.

Suppose we give an ad hoc name to each number, with the sole restriction that it be one not already used, having no connection with the names of other numbers. How shall we then arrange the numbers? In what sequence? How

In this Egyptian funerary painting, six scribes watch four workers. The former count and record as the latter measure grain and move it from one pile to another. The Egyptian empires, known for their administrative complexity and sophistication, needed accountants. Seated on the higher pile, the chief scribe calculates with his fingers and tells the scribes facing him the results of his computations. These are recorded on tablets which will be transferred to papyri for the pharaoh's archives.

One Hundred Thousand Billion Poems

can we calculate? Such a nomination without rules would quickly make any use of numbers impractical.

For this reason, numbers require systematically assigned names that indicate something about their quantity and value. Instead of inventing an entirely new name for each new number, names are based on the names of smaller numbers. For example, the number represented by the name *eighteen* indicates that it is the sum of the numbers represented by *eight* and *ten*.

In a book published in 1961, *Cent mille milliards de poèmes* (in English, *One Hundred Thousand Billion Poems*), the experimental French poet and novelist Raymond Queneau (1903–76) introduced ten sonnets, each with fourteen lines, written in such a way that the reader could choose to replace each line with the corresponding line from one of the nine other poems. The reader could thus compose 10^{14} different poems, that is, 100,000,000,000,000 poems, all adhering to the immutable rules of the sonnet.

Words to express numbers

The words responsible for verbalizing numbers are called *numerals.* From the same number you can make a numeral (I have six), or an adjective (it is six hours long), or a substantive (set the table for six). In addition, there are multiples—double, triple—as well as fractions—one half, one third, one quarter. We have already seen that the cardinals indicate quantity (one, two, ten, one thousand), while the ordinals indicate sequence (first, second, tenth, one thousandth).

In English and European languages the list of numerals starts with the *units,* or *digits*—that is, the single numbers: one, two, three, four, five, six, seven, eight, nine, zero. The first series of ten numbers that are compounds of ten also have individual names: ten, eleven, twelve, thirteen, fourteen, fifteen, sixteen, seventeen, eighteen, nineteen. After that, only certain multiples of ten are named: twenty, thirty, forty, fifty, sixty, up to one hundred. From this point on, only the large round numbers are named, and here we find that naming systems differ in different countries for certain large numbers. A chart on page 131 explains this.

Using fewer than thirty separate names, one can name numbers with 304 digits—and with British usage, 601! This is remarkably efficient, but still just a drop in the ocean of numbers, which measure no more than a *millisecond* to a *gigayear* in the face of eternity.

In Mesopotamia clay was used as a writing surface. An accounting tablet dating from 2400 BC reveals the nail and chevron shapes that formed the digits of this civilization's numeration system.

Written numerations

The intellectual development of the numerical idea follows this evolution: from quantities to numbers and then from numbers to figures. In the early Babylonian land of Sumer, in about 3300 BC, writing was invented, perhaps in response to the need of an imperial culture to manage extensive lands, herds, crops, and populations. Accounting became more and more complex and soon necessitated written records. Thus was born the written

representation of numbers, which seems to have occurred at the same time as the creation of the first cuneiform. The first written numeration was Sumerian and in the earliest clay tablets bearing written language, numbers also appear.

Over the millennia the forms of written language changed a great deal, shifting from pictograms and ideograms to phonograms, but numerical symbols never underwent this transformation. Instead, they became specific symbols reserved exclusively for the representation of numbers. These symbols became digits.

The caste of accountant-scribes (shown above in a fresco from Til Barsip in Syria) managed the wealth of the Assyrian Empire. This conferred on it a power equal to that of the warrior and priest castes. The scribes drew up numerical tables and wrote numerous arithmetic treatises offering problems and their solutions, and were also responsible for teaching. The Assyrian symbol at left, called a *nail*, represents the unit, or one, while the *chevron* shape beside it stands for ten. The two numbers shown are thus 2 and 20.

What is a digit?

Digits are specific numbers that have the task of representing all the numbers. Each is designated by a specific symbol. In modern usage we employ a set of digits that are called *Arabic numerals:* 1, 2, 3, 4, 5, 6, 7, 8, 9, 0. Other written forms include the nail and the stamp of Sumerian numeration, "lotus flower" and "frog" in Egyptian numeration, and the point, line, or glyph in Maya numeration.

Some civilizations—the Greeks, the Romans, the Hebrews—did not create special symbols for numbers; instead, they assigned a specific letter to each digit. These are called *alphabetical numeration systems.* The Hebrew *aleph* (א) is 1, *beth* (ב) is 2, *gimel* (ג) is 3, just as the Greek *alpha* (α) is 1, *beta* (β) is 2, *gamma* (γ) is 3.

In the writing of numbers, digits perform the same function as the letters of the alphabet in the writing of words. Conceptually, the invention of numbers preceded digits just as words preceded letters. And just as *a* is simultaneously a letter and a word in the English language, *4* is both a digit and a number in our numeration system. Numbers that are compounds of more than one digit, such as *13,* are never classified as digits, however.

The funerary rites of the ancient Egyptians provided the dead with food and drink, shown above on a table of offerings. The full count of offerings and foods was depicted to assure the deceased of nourishment in symbolic form even when the offerings of the rite itself were gone. This scene is from the tomb of the Princess Nefertyabet at Giza, from about 2600 BC.

A numerical language

Written numerations constitute a separate language that coexists with the language of words. Each has its own vocabulary and its own syntax. The syntax of numerations consists of operations to build up groups of digits in certain arrangements to represent numbers.

What are the digits? How should they be arranged?

Like the majority of great civilizations, ancient Egypt had several numeral systems. The oldest is hieroglyphic notation, dating to the 3d millennium BC. Decimal and additive, it uses signs for the first six powers of ten; that is, it can represent numbers up to one million. *One* is represented by a vertical line, *ten* by a basket handle, *hundred* by a coiled rope, *thousand* by a lotus flower blooming on its stalk, *ten thousand* by a raised hand with a curved little finger, *hundred thousand* by a bull's head, and *million* by a god with arms raised to the sky. The bas-relief at left shows a tally of the quantities of offerings left for the buried princess.

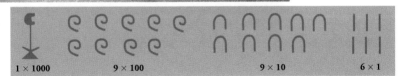

| 1×1000 | 9×100 | 9×10 | 6×1 |

How do we represent a given number? Conversely, how do we "decode" a written numeration and determine what number it represents?

The rules governing numerical syntax are strict: the same written numeration must never represent two different numbers. Yet all of the numeration systems have some ambiguities, so that context is important in the reading of numbers, as it is in words.

The scribe of Giza who carved the wall of Princess Nefertyabet's tomb would have written the year 1996 like this.

IIIIIIIVVVIVIIVIIIIXXXIXIILCDM

Making much out of very little

The spatial arrangement of digits is linear: the symbols are set out on a horizontal or vertical line. Most systems use a horizontal line, giving the impression of reading text. A written numeration is characterized by the fixed meaning of the digits and by the operations used to create numbers out of sequences of digits. The representation of numbers in written numerations, as in spoken numerations, requires an efficient, systematic method. Names must be assigned and used according to universal

.٩٨٧٦٥٤٣٢١

principles applicable to all numbers. And each name must be descriptive; it must tell us about the number it represents: tell me your name and I will tell you how many you are.

A numeration is a system that can make do with very little. Its efficiency depends on one simple organizing principle: the concept of the base.

1234567890

Counting in bundles: the principle of the base

The base system is the means by which numerations are able to use a small number of words (spoken numerations) or symbols (written numerations) to represent a vast variety of numbers. It allows them to escape being pure enumerations in which each

Left, from top to bottom: Roman, Arabic, and Indo-Arabic numerals.

"It is…preferable, we believe, that our students, even those with degrees, spend the holidays going to school than to taverns, and argue with their tongues rather than fight each other with their daggers. Thus we would like the graduates of our school…to discuss and read computation and the other branches of mathematics for free, for the love of God."
Ordinance of 1393, Vienna

"The first thing to be done was to express all possible numbers in a simple way.… This was begun by expressing the first nine numbers with particular symbols; once these had been established, someone had the good idea of giving these characters, besides their absolute value, a value dependent on their position. The character 1, which represents the unit, expresses a unit of the second order, or ten, when it moves one position to the left."
Pierre-Simon, Marquis de Laplace (1749–1827), first course at the Ecole Normale, 1795

number is merely a sum of ones: $n = 1 + 1 + 1 + 1 + 1 + 1 + 1 + 1 + \ldots + 1$ (n times).

Just what is the base system? Instead of counting and naming an endless string of units, we count in bundles or sets. These bundles establish a hierarchy in the sequence of numbers, defining units of the first order, of the second, and so on. A numerical base is the number of units in a bundle. Within a given base, the bundles of each order all have the same number of units.

Medieval universities apparently long tolerated the private teaching of mathematics without incorporating it into the standard course of study. A Roman manuscript of the 13th century shows a teacher and his students, including the Child Jesus.

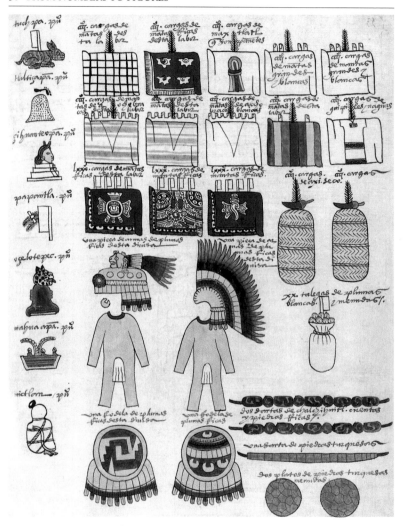

In principle, a base can be formed of any number,
but in fact only a handful have been put to practical use.
The commonest base is the one we employ in the West
today: the decimal base, or base 10. In the base 10 system
we name digits one through nine, plus zero, to form

the first order or bundle. After nine we reuse the names in different forms to indicate the second order, or *power* (the teens), the third power (the twenties), the fourth power (the hundreds), etc. The Sumerians had a sexagesimal base (60), the Maya a vigesimal base (20). Duodecimal (12), quinary (5), and binary (2) bases also serve many purposes—base 2 being the foundation of computer calculation. Base 10 is a convenient and logical choice for many reasons: base 5 is too small and base 20 too big for common usage. Is it mere coincidence that base 10 corresponds to the number of fingers we have?

Additive and hybrid numeration systems

The different numeration systems have been combined in all sorts of ways throughout history. The Egyptians, the Chinese, and the Greeks all had three numeration forms, the Maya two, and the Indians four. The Aztec, Ethiopians, Hebrews, and

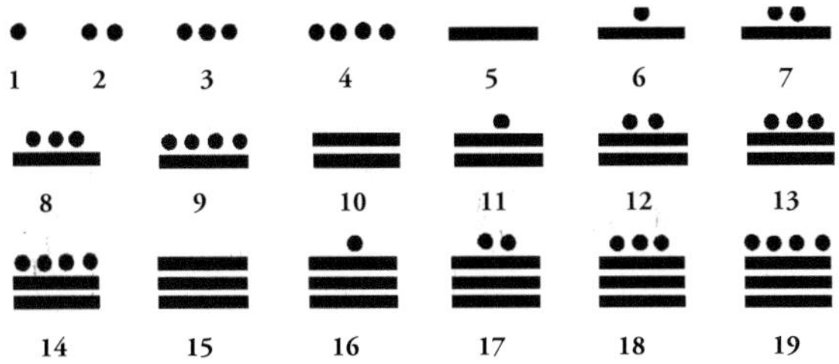

1	2	3	4	5	6	7
8	9	10	11	12	13
14	15	16	17	18	19

Romans all had their versions.

Historians of science, among them Geneviève Guitel and Georges Ifrah, have identified three kinds of numeration systems: additive numerations, hybrid numerations, and positional numerations. These three classifications are distinguished by the sort of arithmetic operations each uses to compose numbers from digits.

In *additive numeration systems,* addition is the only operation employed in the formation of numbers. A number is created by the juxtaposition of symbols; its

Opposite: this sheet from the 16th-century Codex Mendoza adds up the tribute paid by seven Aztec villages to the Spanish overlords of Mexico. In Aztec numeration, vigesimal (base 20), like that of the Maya, and additive, the *unit* is represented by a dot, *20* by a hatchet, *400* (that is, 20 × 20) by a feather, and *8,000* (20 × 20 × 20) by a kind of purse. Four hatchets set on a cape represent 4 × 20 capes; a feather stuck in a bale of dried peppers indicates 400 bales, and so on.

The Maya found the simplest way to represent numbers was with a system that used the dot (equal to 1), the bar (equal to 5), and a zero. The first nineteen Maya numbers are pictured at left. These numerical symbols are also found in the pages of the 13th-century(?) Dresden Codex (overleaf), one of four Maya manuscripts that have been preserved.

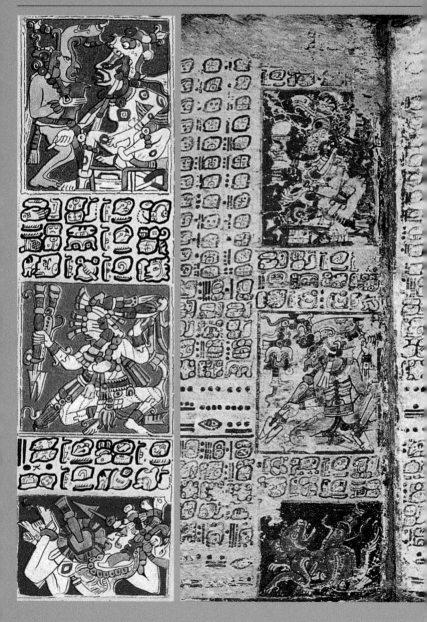

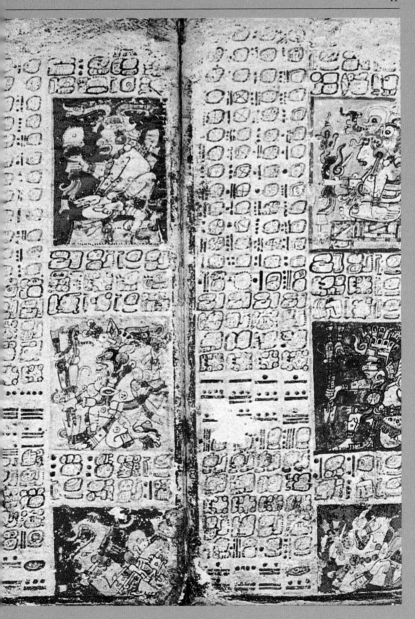

value is equal to the sum of the values of the symbols. Each cluster of symbols is repeated as many times as required; thus, *two hundred* is conceptualized as *one hundred + one hundred*, and is represented by the notation two times of the digit representing *one hundred*. The Roman numeral CCLVI, where *C* represents one hundred, *L* represents fifty, *V* represents five, and *I* represents one, is an example of this. (The number is two hundred fifty-six; we would write it as 256.)

Hybrid numeration systems (such as our own) exploit both addition and multiplication. Addition counts up the contributions of successive powers, while within each power multiplication is used in the following way: *two hundred* is conceptualized as *two* times *hundred* and is represented by *two* followed by *hundred*. This juxtaposition is, in effect, a multiplication. Thus, for example, 243 is two times one hundred plus four times ten plus three times one.

Naming numbers

In these two types of numeration, the different base powers come into play explicitly in writing. Each power is represented by a figure. The drawback with both systems is that new figures have to be added to represent higher numbers. New names would constantly have to be invented for each power. Imagine an alphabet that got bigger each time a new word was coined and had to be written. The advantage of an alphabet is

oo. **Quingenta millia.**

ꓛꓛ.**1000000, Decies
cētena millia.**

**tur ultra decies centena
lunt, duplicant rotas: ut**

that once it is set, it withstands time and invention; it stays the same in the face of any number of new words. This does not happen with the numeration systems described above.

The numbers that they are supposed to represent are unlimited in quantity, while the systems themselves have a limited capacity.

The answer to this quandary is the system known as *positional numeration,* or *positional notation.* How is it that positional numeration systems are able to accommodate an unlimited capacity?

Roman notation is not alphabetic, though it seems to be. The seven numerical symbols—I, V, X, L, C, D, M (1, 5, 10, 50, 100, 500, 1,000)—were not originally letters of the Latin alphabet. The Roman Empire's numerical system is wholly inadequate to work with large numbers, so that the simplest calculation is arduous. Multiply LVII by XXXVIII and you get MMCLXVI (57 × 38 = 2,166)! Above: a bronze Gallo-Roman engraved calendar, late 1st century AD; left: Roman numeral calculations.

We are so accustomed to making computations by writing them down that the process seems to us as old as civilization itself. Yet the use of writing for calculation turns out to be a late and unusual practice in human history. Computing by writing could only be accomplished with the invention of positional notation, using zero. The concept of zero was a new and radical invention, one that changed the world, so that ten symbols represent all the numbers in the universe.

CHAPTER 3
POSITIONAL NOTATION

"A donkey on the highest step has a higher value than a lion on the lowest." This is the principle of place value: the 1 in 1,000 has a higher value than any of the three 9s in 999. Right: a 16th-century mathematician; opposite: a detail of his work.

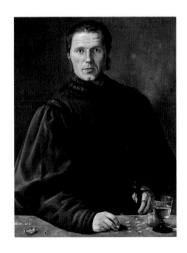

The principle of position: make the place count

In most numeration systems the *value* of a digit has no relation to the *position* it occupies in the writing of the number. The *I* of Roman numerals, for example, is equal to *one* no matter where it is written in a sequence of numerals. *M* is always equal to *one thousand*. Thus, one thousand and one is written MI.

Positional notation breaks with this principle and rises to a higher level of abstract thinking. It maintains that the value of a digit is not constant, but varies according to the position it occupies in the number. Here, in the true sense of the word, the place "counts"; it has a particular value, which is combined with the value of the digit itself to make the value of the number.

The necessity of zero

In most positional numeration systems the places are written horizontally, like letters in a word. Each place corresponds to a power of the base. Because the place itself has a fixed value (the places for ones, tens, hundreds), that value need not be explicitly represented by a digit when a number is written. Thus, digits that mean *ten, hundred, thousand,* and so on are no longer needed; there are now only quantities of tens, hundreds, thousands. The only symbols, or *notational digits,* that remain in use are those representing the units.

The principle of position was originally used in figural numerations and was later adopted in written numerations. To pass from the representation of a number on a calculating board or abacus to its written form is only a matter of converting the physical objects— the rods or columns of the abacus and the beads, counters, or pebbles threaded on them—to the more abstract grid of placements on a page, marked with

When computations are made with a mechanical device, such as an abacus, there is no need to set down the operations, or record their sequence. With written computations, it becomes a necessity. What is the best spatial arrangement for the different stages of a calculation? The manner of setting down such written operations has evolved greatly over the centuries. Below: a division style in use in Europe in the 16th century, called *galley division,* from the name of a Venetian monk, or possibly from its form,

which resembles a ship. Venetian students turned their work into a drawing after they had completed it. This method had been used in the 9th century by the Arab mathematician al-Khwārizmī.

written symbols. There is only one element that an abacus has that our written repertory of symbols lacks: the rod with no beads on it, indicating an empty column. Thus is the final digital symbol born: the *zero*.

Positional numeration systems are the only ones that require the zero. For example, the tens place must stand even when the column that represents it has no occupant. A symbol must be placed in that column

Sacrobosco, a 13th-century English mathematician, was author of the best-seller *General Algorism* (above: a miniature from it). He played an important part in the dissemination of Arab numerals in the West.

to tell us both that it has no digits in it and that the digits in the next column are in the hundreds place.

In the number 1,001 the tens and hundreds do not count —that is, they have no content; the second and third places are marked by 0. And though the two other digits are identical in form, we can identify their differing values by their positions: the first (on the right) is equal to one; the second, three places to the left, is equal to one thousand.

The power of zero

The great elasticity of positional notation is due to several elements: it has the zero, it uses the decimal base, and its digits function independently of one another. That is, its way of writing them is such that none can be misread as the juxtaposition of several other numbers; to put it another way, its numbers cannot be broken down. It is this independence of the digits in relation to one another that excludes all ambiguity in their reading (often a problem for the other numeration systems) and makes this system so flexible and adaptable.

A single strategy of an astonishing simplicity governs this system: the *principle of position*. This principle has a rather democratic quality in that the digits, placed one after another in a line, do not adhere to any rule of precedence, which would circumscribe their use. All places are permitted to all—including zero. (A unique exception is the number 0123…, in which the zero is not necessary.) From this principle it follows that every group of digits that respects this rule represents a number and only that number, and conversely, that every number is represented by one, and only one, group of digits. That the digits are placed in a horizontal line makes it easy to read them in a natural way.

Another advantage of this numeration system is the link it establishes between the length of the name and the size of the number. The longer the name, the larger the number. Such a link between the length of the name

The written name of a year (above left) is an example of positional notation. Note the absence of a comma, which is not necessary in this case.

Let's take a look at some basic arithmetic operations. Multiplication proceeds from addition and can be defined as the repetition of additions: $m \times n = n + n + n + n + n + \ldots + n$ (m times). Similarly, *raising to a power* is defined as the repetition of multiplications: $n^m = n \times n \times n \times n \times \ldots \times n$ (m times); m is the *exponent*. In mathematical notation a power is written with the exponent as a small, raised number. 1 followed by n zeros is thus written as 10^n and expressed in words as "ten to the nth power." It is equal to 10 multiplied by itself n times. Thus, $100 = 10^2 = 10 \times 10$; $1{,}000 = 10^3 = 10 \times 10 \times 10$; and so on. Note a few traits of this system that may at first seem odd: $10^1 = 10$; $10^{n+1} = 10^n \times 10$; and $10^0 = 1$.

Two artists (left and below) rethink the position of some digits.

and its value makes comparison extremely simple. For example, as 1,001 has more digits than 888, it follows that 1,001 is larger than 888. Compare this with Roman numerals, where 1,001 is written MI, with two digits, while 888, the smaller number, is written DCCCLXXXVIII, with twelve digits.

All the numbers in the world

Numeration has thus developed a one-to-one correspondence between numbers and their names. That is why it is so hard for us today to understand the difference between a number and its name in positional notation. For us, what is the number represented by *1,001* if not the number itself? The system of nomination-representation invented in 5th-century India was the most developed form of positional notation. It has the same unlimited capability as the totality of numbers it is responsible for naming.

Furthermore, the quality of Indian numerical represen-

The number 109,305 appears in the 12th-century Indian manuscript known as the *Bakshali,* a page of which is pictured below. Zero is represented not by a circle but by a point, the *bindu.*

tation has attributes that no alphabetical language can boast. Every group of digits is the name of a certain number. Not every group of letters of the alphabet forms a word —many are meaningless gibberish. The written group of

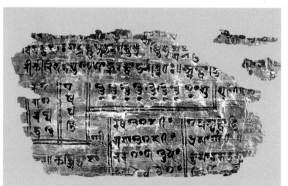

letters *eejkwxjj* is not a word in English, French, or any other language; it represents nothing, but any group of numerals means a specific number.

Indian positional numeration has the capacity for unlimited representation: ten digits alone—the number of fingers on our two hands—can represent all the numbers in the world.

The invention of Indian positional notation

Why is this wonderful system called *Indian*? The figures for the digits, from one to nine, were invented in India more than two thousand years ago. They appear in inscriptions from the 3d century BC. But the principle of place value had not yet been applied at that time, nor had the zero—the symbol for *nothing,* or *a quantity of none*—been invented. It was the combination of these two concepts, the idea of place value and the sign to represent a value of nothing, that together made a number system flexible and sophisticated enough to lead to modern mathematics. This combination occurred in India in the 5th century AD, and passed via the Arabs to Europe over the following centuries.

In 458 there appeared an Indian treatise on cosmology, written in Sanskrit: the *Lokavibhaga,* or *The Parts of the Universe.* In it the number fourteen million two hundred and thirty-six thousand seven hundred thirteen was written, using the place-value system and requiring only eight digits: 14,236,713. In the text the digits were written entirely in letters and from right to left: "three, one,

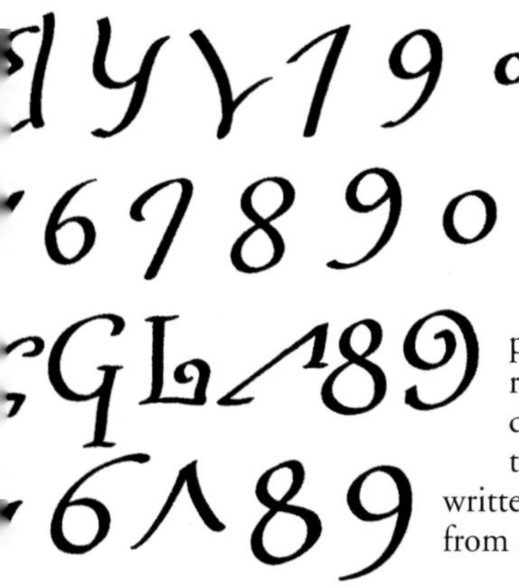

The power of the graphic form of the Indian digits—and the conceptual thinking that went into their creation—resides in the fact that each digit is independent of the others. That is, neither the two nor the three nor the four nor any other of the first order of numerals consists of groups of ones. This is its great advantage over the other three ancient positional numeration systems. The fact that the Indian system is so specific confers total autonomy on each of the ten symbols and frees written numeration from all ambiguity. Emigrating to the Near East, Indian digits gradually altered as they spread through the Arab world. Their graphic evolution is depicted opposite. In the countries of Arabic North Africa and in Spain, the numerals took a form quite different from the original Indian digits. These later Gobar numerals made their way to Europe and lent their forms—and the name *Arabic*—to those in current use in the West today. Modern Arabic numerals in the Arab world remain somewhat different.

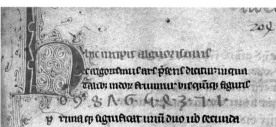

seven, six, three, two, four, one." The word *sunya,* void,
which represents zero, also appeared. This is the earliest
known document to use what we now know as Indian
positional notation.

The dissemination of positional notation by the Arabs

Some three centuries later, in 773, an Indian
ambassador came to Baghdad bearing treasures:
a knowledge of digits and computation. The
Caliph al-Mansūr (c. 709–775), ruler of the city,
was a great patron of learning; he and the
Arab scholars at his court immediately
recognized the great value of this gift, and
began to foster its study.

In Baghdad in the first decades of
the 9th century the great Arab mathematician Muhammad ibn Mūsā al-
Khwārizmī (c. 780–c. 850) wrote the
first work to present the new
knowledge in Arabic, *The Book of
Addition and Subtraction by Indian
Methods.* This book was the means by
which Indian computation arrived in
Europe. Immensely influential and
translated into Latin many times,
beginning in the 12th century, its
renown was such that the method of
computation it presented became
known as *algorism,* from Algorismus,
the Latin name of al-Khwārizmī. A
Latin poem, *Carmen de algorismo,*
written around 1200, states: "We
call the present art by which we use

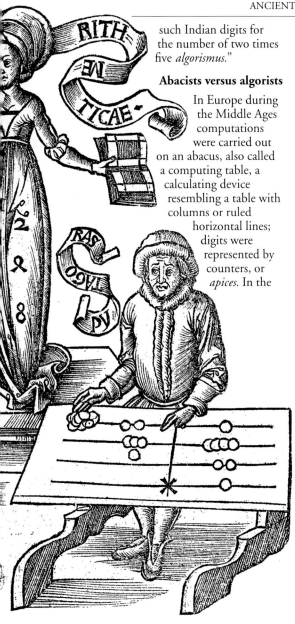

such Indian digits for the number of two times five *algorismus*."

Abacists versus algorists

In Europe during the Middle Ages computations were carried out on an abacus, also called a computing table, a calculating device resembling a table with columns or ruled horizontal lines; digits were represented by counters, or *apices*. In the

One writes, the other doesn't. The algorist triumphs over the abacist. This engraving (left) from the early 16th century depicts the victory of written computation over that performed with counters. In the background, Lady Arithmetic, wearing a dress strewn with numerals, makes it clear which side she favors. Founded on the nine digits and zero, algorism made carrying out the four operations on whole numbers (addition, subtraction, multiplication, and division) faster and more accurate.

Opposite: the *Carmen de algorismo*, or *Poem on Algorism*, by Alexandre de Villedieu, played a major role in the spread of the new arithmetic in Europe, especially in the French universities, among the earliest to teach mathematics. The ten digits are depicted in its opening, which reads from right to left, as Arabic writing is read.

Carmen de Algorismo, for the first time in Europe, zero was considered a digit. Raoul de Laon, skilled in the use of the abacus, had the idea of placing in the empty columns a character he called a *sipos,* or counter. This counter, soon replaced by the symbol 0, rendered unnecessary the columns of the abacus. From the 12th century on, this type of abacus was progressively replaced by the dust board as a tool of calculations.

This development did not come about without a struggle between those who, evoking the ancient Greek mathematician Pythagoras (c. 580–c. 500 BC), championed the abacus and those who became masters of algorism, the new Arabic number system. In this competition between the Ancients and Moderns, the former often saw themselves as the keepers of the secrets of the art of computation and the defenders of the privileges of the guild of professional calculators, with interests paralleling those of the Christian church. The introduction of the new system indisputably marked the democratization of computation: its simplicity and lack of mystery made its widespread use possible. Computation was no longer an esoteric art practiced within the limited circles of specialists.

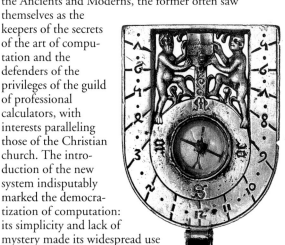

In medieval European culture sundials enjoyed a special status. They displayed all the numbers attributed to the different hours of the day, engraved in stone or metal. The appearance of Indo-Arabic numerals, replacing Roman numerals, on sundials played a part in their popularization. A portable sundial from the mid-15th century, left, is paired with a compass.

The written digits that we use today, and that we call Arabic numerals, came not from the Arabic Near East but from the western Arabs of Moorish Spain. They were called the Gobar numerals. They traveled an extraordinarily long road: from India to the Near East to Arabia to North Africa to Spain. The journey took eight hundred years.

Gradually, the Indian origin of positional notation

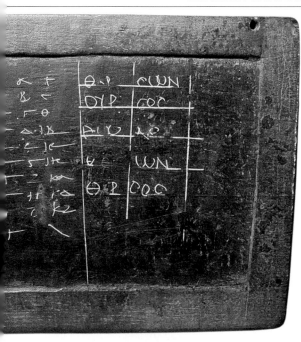

Like Hebrew notation, the numeration system developed in the Greek world in the 4th century BC was alphabetic. The digits were represented by capital letters: three sequences of nine letters for the units, tens, and hundreds, utilizing all twenty-four letters of the Greek alphabet, plus three new signs to complete the notation. This system was decimal and additive. Thus, 1,789 was represented as:

῾Α   Ψ   Π   Θ

1,000 + 700 + 80 + 9. The number 1,000 was obtained by the addition of a symbol (an apostrophe) to the symbol representing 1, an A, or *alpha*. However, Greek mathematicians developed more powerful numerations. In his treatise the *Arenarius* Archimedes (287–212 BC) created a system that could comprehend numbers up to 1 followed by eighty million billion digits. Greek mathematics made a clear distinction among the arts of computation, logistics, and arithmetic, the pure study of number. Above left: a Greek tablet from the early 1st century AD, showing multiplication.

was forgotten in the West, and only its immediate source was remembered. Thus, the Indian digits became known as Arabic numerals and the zero was thought to be an Arabic invention. The Arab calculators were the first to promote and disseminate the Indian system widely; Arabic tradition continues to credit India as its true source.

Computing with ordinary tools

The Indian calculators of the 5th century and their Arab successors wrote their figures on the ground or on dust boards. They carried dust or flour in a small bag and traced the digits with a finger, a stick, or a pointed knife. Later, they began to use a tablet coated with wax, which was incised with a stylus; the wax could be rubbed smooth and reused. This was replaced by a slate marked with chalk, and still later by ink, pen, and paper.

Until the Indian system came into use, the recording of numbers with the help of a writing system had always been considered quite separate from the development of the writing and recording devices that made computation possible. Indian positional notation abolished the distance between writing and computation. Specialized calculating devices such as the abacus were no longer needed; from this point on, operations could be carried out directly with a pen and a sheet of papyrus, parchment, or paper, using just the numbers themselves. This gave rise to *computational writing,* which could serve for the most complex computations.

Can the present system be improved? "Our place-value system constitutes a perfect system," writes the historian Georges Ifrah. "The invention of our current numeration system forms the terminal stage in the history of numerical notation: after it was achieved, no further discovery

The numerals that represent the numbers from 1 to 9 in three types of Chinese notation are shown in the boxes below. In each box the large symbol displays the oldest notation (1450 BC); below it at left are pictured smaller commercial digits; scientific numerals are at right (2d century BC). These are a numeration in base 10 using only two symbols, a vertical and a horizontal line. Bottom: a Russian abacus.

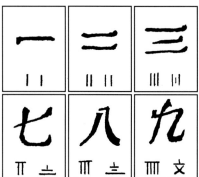

was possible in this area." These unsurpassable qualities have guaranteed its universality and its persistence.

Together with the decimal metric system, Indian positional notation is used today by nearly all the peoples of the world.

The three other positional numeration systems

Aside from Indian notation, there are three other instances of place-value systems that developed independently. These arose in Babylon

Moulfrat ars munea que Untus pollit babere
Cyphca permmien que lit proportio rerum

at the beginning of the 2d millennium BC, in China in the 1st century BC, and in the Maya Empire during a period roughly between the 5th and 9th centuries AD. All three show a great degree of intellectual sophistication, as one would expect of three highly developed cultures, but all share the same shortcoming: their representations of the units of the first order are not independent. Two, for example, is not a specific digit but the doubling of one. The Maya place-value system, with a base of 20, instead of presenting nineteen different symbols offers only three, those representing one, five, and zero. Similarly, Sumerian

In the Renaissance the manipulation of numbers and the practice of arithmetic were signs of advanced learning; those who knew how to multiply and divide were guaranteed a professional career. In this 16th-century tapestry Lady Arithmetic teaches the new calculation to gilded youth.

notation, with a base of 60, deploys not fifty-nine separate symbols but only two, to represent one and ten. The same limitation is true of the Chinese decimal system.

Binary notation: ancient and modern

Another system frequently used instead of base-10 positional notation is binary notation, which uses exclusively the digits 0 and 1. Next to pure enumeration, it is the simplest system imaginable. It is both the oldest and most modern of the positional numeration systems. The most ancient version of it is that of the Torres Strait, between Australia and New Guinea. Its inhabitants used a numeration called *urapun-okosa,* characterized by the alternating of ones and twos:

1 = *urapun*
2 = *okosa*
3 = *okosa-urapun*
4 = *okosa-okosa*
5 = *okosa-okosa-urapun*
6 = *okosa-okosa-okosa*
…and so on.

Limited in its ability to express certain kinds of complexity, binary notation is nonetheless the basis for all computer functions. The modern use of the binary system in the coding of numbers gives the computer its extraordinary powers of calculation. The German philosopher and mathematician Gottfried Wilhelm Leibniz (1646–1716) was the first modern thinker to promote this system; in 1703 he announced that instead of the sequence of ten, he had come to prefer what he called the simplest of all progressions, two by two, which served the science of numbers perfectly. He explained that he used no other characters but 0 and 1, and once he reached two, he began over again. For this reason he

wrote *two* as 1 0, and *two times two,* or *four,* as 1 0 0. *Two times four,* or *eight,* was 1 0 0 0, and so on. The length of numbers in binary notation is at least double that of numbers in the decimal system (except for 0 and 1, of course). This makes the binary system impractical for human calculators, but it does not upset computers in the least.

From the computer's point of view, these sequences of 1 and 0 are convenient, for they are easily codified in electric signals: the passage of current expresses 1, its interruption 0. These 1s and 0s are interpreted as sequences of on and off, or yes or no, proliferating at extraordinarily high speeds that permit extremely long numbers to be coded fast.

Above: binary notation interpreted in the modern style. The German philosopher Leibniz (opposite above) drew a medallion (opposite below) extolling binary notation. He wrote: "It is true that as the empty voids and the dismal wilderness belong to zero, so the spirit of God and His light belong to the all-powerful One."

60

What do we know of the numbers 1, 2, 3…, which appear so obvious that we call them *natural*? The limitless sequence of *integers,* seemingly so familiar, is full of mysteries. Natural numbers have all sorts of patterns and rules, some of them unexpectedly regular, and many remarkable properties. Their study constitutes the theory of numbers, called *arithmetic.* The great German mathematician Carl Friedrich Gauss (1777–1855) called arithmetic the "queen of mathematics."

CHAPTER 4
NATURAL NUMBERS

Generations of primary-school students recited multiplication tables and balked at doing division. Do numbers still seem mysterious as they did in past centuries? The game at left, called the Little Mathematician, was used in schools at the beginning of the century. It could carry out the four fundamental operations of arithmetic: addition, subtraction, multiplication, and division. Right: a child of the Victorian era struggles to add 2 + 2.

Investigating natural numbers

Once numbers acquired place value and the zero, their complexity grew rapidly. They developed new attributes and new, more refined definitions. A *set* is any collection of numbers that belong to a defined category. A set can be *finite* (the set of hours in the day) or *infinite* (the set of hours in the future). *Natural numbers,* also called *positive integers,* are the set of numbers, from zero on, obtained by adding 1, first to 0, and then in sequence to each number so obtained, thus: 0, 1, 2, 3, … (the three ellipsis points "…" indicate "and so on to infinity"). This set is conventionally named **N**. Integers are also called *whole numbers*, as distinguished from fractions.

Looking at the set **N** we can see that natural numbers have a beginning—0 is the first natural number—but no end: there is no last natural number. Further, each natural number *n* has an immediate successor, (*n* + 1). We will discover other such properties of this set as we explore its characteristics and functions.

The study of *positive integers*, *negative integers*, and *fractions* belongs to the domain of *arithmetic*, the science of numbers, which analyzes the behavior of various numbers in four operations: addition, subtraction, multiplication, and division. It is among the most difficult of the mathematical disciplines.

Numbers derive their existence

Euclid, the great Greek mathematician of the 3d century BC (rendered, opposite, by the Surrealist painter Max Ernst [1891–1976] in 1945 as a thinking triangle) was author of the *Elements*, the most widely used mathematics textbook of all time. Though known mainly as a geometer, he developed a method of division known as Euclidean division, using the concept of the remainder. Thus, 19 divided by 5 equals 3, with a remainder of 4.

The triangular multiplication table at left and the smaller division tables come from a French manuscript of 1793, the work of a student composing a complete course in arithmetic.

and sequence from addition (1 + 1 + 1 + …). Multiplication begets the place-value system (remember: $243 = 2 \times 10^2 + 4 \times 10 + 3$). Division is an essential investigative tool: we learn more about the natural numbers by testing their *divisibility*—a key word of arithmetic.

Even and odd: the first classification of natural numbers

Why is division so important to the definition of natural numbers? Because the property of being divisible without creating fractions varies among numbers—some are more divisible than others. The study of divisors is one of the principal modes of classifying the integers. How does each react to division? Is it divisible by this number or that? Is it *highly* or *hardly* divisible? And so on.

Division by 2, the simplest operation of division, gives us the first classification of integers. *Even numbers* are those divisible by 2; those that are not are the *odd numbers*. An even number can be divided into two halves, both of which are integers. The ancient Greek mathematicians called the Pythagoreans (6th–5th century BC) were the first to verify this distinction and to establish a general rule from it. The steady alternation of even and odd marks the sequence of the natural numbers.

To signify that even numbers are doubles, mathematicians note them as $2n$, n being any of the natural numbers; it follows that the odd numbers are written $2n + 1$.

What operations preserve parity? The *sum* of two even numbers is even, but so is that of two odd numbers. Thus, addition does not preserve parity. On the other hand, the *product* of two even numbers is even, while that of two odd numbers is odd. Therefore, multiplication preserves parity. And squaring? Squaring is a form of multiplication. The square of an even number is even, that of an odd number is odd. So squaring preserves parity.

Most people make a distinction between good and bad, lucky and unlucky numbers. The artist Charles Demuth (1883–1936), who titled his painting, opposite, *I Saw the Figure 5 in Gold*, evidently liked that numeral. In a North African tradition the five-fingered hand of Fatima wards away bad luck. In some countries the number 13 is unlucky; in others it is 17.

Place your bets, ladies and gentlemen! Play the numbers! You could win a lot! How many games are based on numbers, from roulette to the lottery?

Prime numbers: the second classification

The second stage in classifying numbers is to identify those that cannot be divided except by themselves or 1. (A number is always equal to the product of itself by 1: $n = n \times 1$.) Such natural numbers are called *prime numbers* because they are not the product of any other integer but themselves and 1. This group includes the numbers 2, 3, 5, 7, 11, 13, 17, and an infinite list of others.

Every integer is either a prime number or the product of prime numbers. To identify a number by its primes is called *decomposition into prime factors.* Each number has one and only one formula of decomposition, unique to it—a fact of great importance. Since every decomposition into prime factors is unique to a particular integer, the collection of its prime divisors can be considered its signature. For example, [7, 11, 17, 23] is the signature of 30,107, as this is divisible by these four prime numbers and only by them: $30{,}107 = 7 \times 11 \times 17 \times 23$.

Prime numbers play the role of generators: they alone can create all of the natural numbers. Since they function as points of support on which the entire architecture of integers rests, knowing them is essential in mathematics. How can we recognize them? How many are there? How are they distributed among the other, non-prime numbers?

The larger a number, the greater the quantity of numbers

smaller than it—and therefore the greater the amount of potential divisors. It follows that the larger a number, the smaller its chance of being a prime number. Higher numbers have a lower density of primes: they become

Left: Paul Klee (1879–1940) inserted the prime number 17 into a work of 1926 called *The 17 Wanderers.*

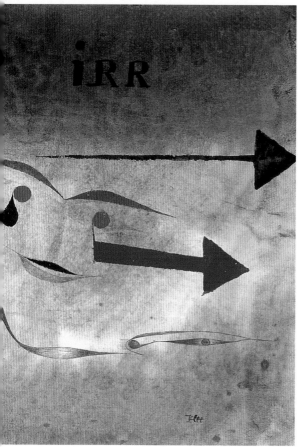

On 8 September 1985, at 7:30 AM, a newsflash from Houston, Texas, alerted the world to the discovery of the largest known prime number: $2^{216,091} - 1$. This is a number with 65,050 digits! Among the prime numbers those of Marin Mersenne (1588–1648), which take the form $M^n = 2^n - 1$ (where n is an integer), are valued highly. Finding them is a popular sport. Between Mersenne's lifetime and the 1960s twenty-two of these were known. The IBM corporation used a Mersenne prime in part of its letterhead (opposite below). A special postmark was created (below) for the discovery of the twenty-third, $2^{11,213} - 1$, in 1963. The twenty-fourth emerged eight years later and seven years after that, the twenty-fifth was discovered by

$$2^{11213} - 1$$
IS PRIME

rarer the higher they get. Do they reach a point at which they disappear? No. No matter how high the numbers become, there will always be prime numbers: there is no prime number that is the largest of all. If mathematicians have a cherished dream, it is to invent a device capable of producing all of the prime numbers.

two eighteen-year-olds, Laura Nickel and Curt Noll. Overleaf: the graphic designer Erté (Romain de Tirtoff, 1892–1990) interprets 3 and 7.

Prime numbers can never be closer than a difference of two. *Prime pairs* are two prime numbers that differ by two. Examples are 17 and 19; 29 and 31; and 1,000,000,061 and 1,000,000,063. It is very surprising that there should be two such large prime pairs when the density of primes at such high numbers is so low. One might suppose that after passing a certain threshold there would no longer be any prime pairs.

Astronomy is a science that has always relied on mathematics. Different methods of measuring the distances of stars and planets in use in 16th-century France are shown here: the quadrant (left), the *carré géométrique* and *nocturlabe* (opposite, far right), and the *bâton de Jacob* (center). Above: an antique clock divides the day into sets of twelve units.

60: the ace of divisibility

The more divisible a number is, the more parts it has. And the more parts it has, the more useful it proves in certain situations. The number 60 is divisible by many different numbers. This explains the secret of 60 and its employment in numerous areas, such as the ancient Sumerian sexagesimal base system. Is it because 60 is highly divisible that the hour has been divided into 60 minutes, and the minute into 60 seconds? Look at the list of its twelve divisors: 1, 2, 3, 4, 5, 6, 10, 12, 15, 20, 30, 60. Compare this with the larger number 100, which has only nine divisors: 1, 2, 4, 5, 10, 20, 25, 50, 100. And 60 carries in its wake its *factor* 12, which has six divisors (1, 2, 3, 4, 6, 12), while 10 has only four (1, 2, 5, 10). This explains why the duodecimal system (using base 12) occasionally takes precedence over the decimal system—in our own twelve-hour day, for example. Until the 18th century many countries divided the day between sunrise and sunset into 12 hours, as a result creating hours of different length during the course of the year: short hours in winter, long hours in summer.

Another relative of 60, its multiple 360, provides the degrees for the measure of angles and the arc of the circle. One of the consequences of this is that the right angle has 90 degrees. For a long time, even after the adoption of the decimal system, many calculations were made with sexagesimal fractions (sixtieths).

Perfect numbers, amicable numbers

Measuring the relative divisibility of a natural number can also be done by looking at the sum of its divisors. Compare a natural number with the sum of all its divisors except itself. If the sum of its own parts is greater than the whole, the number is called *abundant*. Such numbers include 12, 18, and 20.

The angle at which an object is seen is a function of the distance separating object and viewer. If we know the distance and the angle, we can measure the object's size. This is one of the practical purposes of *trigonometry*, a science whose name means "measurement of triangles." The mathematician François Viète (1540–1603), author of a treatise on the angles and the *Canon mathematicus seu ad triangula* (1579), may be considered its founder. The trigonometric functions—sine, cosine, tangent, and cotangent—yield the description of an angle by the numbers that characterize it. These numbers have been calculated and set in tables.

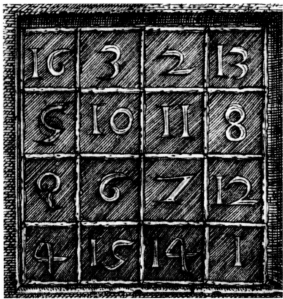

In his allegorical engraving *Melencolia* (below; detail at left) the Renaissance artist Albrecht Dürer (1471–1520) inserted a square of four numbers per side, for a total of sixteen, containing ten identical sums: the four horizontal sums, the four vertical sums, and the two diagonal sums all equal 34. The source of this square is a treatise on magic squares by Manuel Moschopoulos, a Byzantine scholar of the 13th or 14th century; in Dürer's time it was considered a charm against the plague. A lover of science and games of wit, Dürer set the date of the work—1514—in the two central squares of the last line.

For example, 12 is less than the sum of its own parts (12 < 1 + 2 + 3 + 4 + 6 = 16). If the sum of its parts is less than the whole, the number is called *deficient*. Such numbers include 4, 8, 9, and 10. For example, 10 is greater than the sum of its own parts (10 > 1 + 2 + 5 = 8). When the sum of its parts is equal to the number, it is called *perfect*. Such numbers include 6 (6 = 1 + 2 + 3) and 28 (28 = 1 + 2 + 4 + 7 + 14).

According to legend, somebody once asked Pythagoras, "What is a friend?" He answered, "That which is my other self." When his questioner expressed his puzzlement, he explained, "That which is my other self, as 220 is to 284." And what are 220 and 284 to one another? They share a strong bond in relation to their divisibility: the sum of the divisors of one is equal to the

other. This defines them as two *amicable* numbers. The pair cited by Pythagoras forms the smallest pair of number "friends": the divisors of 220, not including itself, are 1, 2, 4, 5, 10, 11, 20, 22, 44, 55, 110, for a total of 284. The divisors of 284—1, 2, 4, 71, 142—form a total of…none other than 220.

In the 9th century the Arabic mathematician Thābit ibn Qurrah (c. 836–901) described amicable numbers. Above: a 13th-century manuscript of his treatise.

Just what is it that mathematicians do?

The field of mathematics is a universe composed of objects, of properties attributed to these objects, and of the truths about them. Mathematicians construct classifications of objects of the same type and establish connections among different types of objects, or between different properties of the same object. Are two properties equivalent? Does a general rule or principle follow from a collection of properties and functions? Does one idea follow from another? Mathematicians are architects of complex systems.

A mathematical question is formulated as a proposition, expressed as a problem. When a mathematician reaches the point of answering the problem by presenting a convincing proof of the proposition, it then becomes a truth of this mathematical universe. If from this problem the set of living and future mathematicians can establish other truths, it then becomes a *theorem.*

To simple questions, some very complex answers

In mathematics a really good problem is generally one that has a simple formulation, but whose solution turns out to be particularly difficult. Arithmetic has proved to be a fount of good problems. The extreme simplicity of the formulation of its questions often masks the extreme difficulty of their solution.

Certain questions asked many centuries ago have yet to be resolved: is there an infinite number of prime pairs? Is there an infinite number of amicable numbers? Are there any odd perfect numbers?

We have seen that every natural number can be broken down into a product of some prime numbers. But can it be broken down into a given sum of prime numbers— two, three, or four, for example? In 1742 the mathematician Christian Goldbach (1690–1764) sent a letter to his colleague Leonhard Euler (1707–83) in which he asserted, without giving a proof, that "every even number (except 2) is the sum of two prime numbers." For example, $16 = 13 + 3$, or $30 = 23 + 7$. Two and a half centuries later, we still do not know if this assertion is true. And odd numbers? It has been demonstrated that every odd positive integer of a certain size (greater than $3^{14,348,907}$) is the sum of at most three prime numbers.

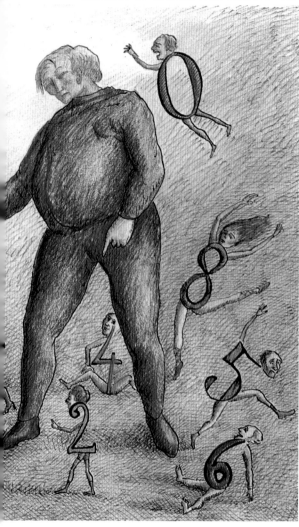

A number tamer? The giant who can achieve such a status has not yet been born. In mathematics, for every conjecture that is resolved another one arises. That is because the structure of positive integers remains mysterious and enormously complex. The tools of mathematics —algebra, analysis, topology, algebraic geometry, and so on— help us to explore this intricate fabric. There is no need to shroud numbers in mystery; one has only to study their troubling architecture to experience all the marvelous confusion one could possibly want.

When the mathematical tribe is convinced of the truth of an assertion, even though no one has been able to prove it, this assertion is awarded the title of *conjecture*. The longer it resists all efforts to prove it, the more renowned it becomes. Anyone who *dis*proves a conjecture is guaranteed lasting fame. All mathematicians, at least once in their lives, have attempted to resolve a celebrated conjecture—which usually withstands the effort.

The successive generations of mathematicians who tackle a conjecture usually go about it gradually, by eating away at it. Seeing that it is not possible to develop a proof in its complete general form, they look for situations in which they may find an answer to specific instances, hoping that from these a general proof may evolve.

Fermat's conjecture

Can a perfect cube be the sum of two perfect cubes? In 1640 the mathematician Pierre de Fermat (1601–65) formulated this question and answered it. In the margin of one of his books he jotted the following note: "There do not exist positive integers x, y, z, n such that $x^n + y^n = z^n$ when $n > 2$."

Fermat is one of the great creative minds of mathematics, an inventor of number theory and differential calculus and a great innovator in probability theory; his proposition has therefore been taken very seriously by subsequent generations of thinkers. But as Fermat did not provide the required proof, his assertion long remained a conjecture, though virtually all

mathematicians became convinced at heart that it was true. Fermat himself, and after him Euler, proved that the proposition holds when $n = 3$ and $n = 4$. Over time other mathematicians gradually enlarged the number of instances in which the statement is true. Then, in 1987, D. Heath Brown established it for "almost all the values of n." In mathematics, however, "almost all" is not all. It took another eight years for the mathematician Andrew Wiles, using powerful computers and complex computation, to demonstrate that the proposition holds for all instances. Since May 1995 it has become correct to call it Fermat's *theorem*.

Opposite: a page from Fermat's *Miscellanea mathematica*. Scribbling his famous conjecture in the blank space of a book, Fermat (opposite, inset) noted: "I have discovered a wonderful proof of this proposition, but it does not fit into the margin of this page." For 350 years frustrated mathematicians tried to fathom what it could be. In 1993 Andrew Wiles displayed the great lines of his proof of Fermat's conjecture (above). Although it covered a thousand pages, it was incomplete. Two years later he furnished the full proof.

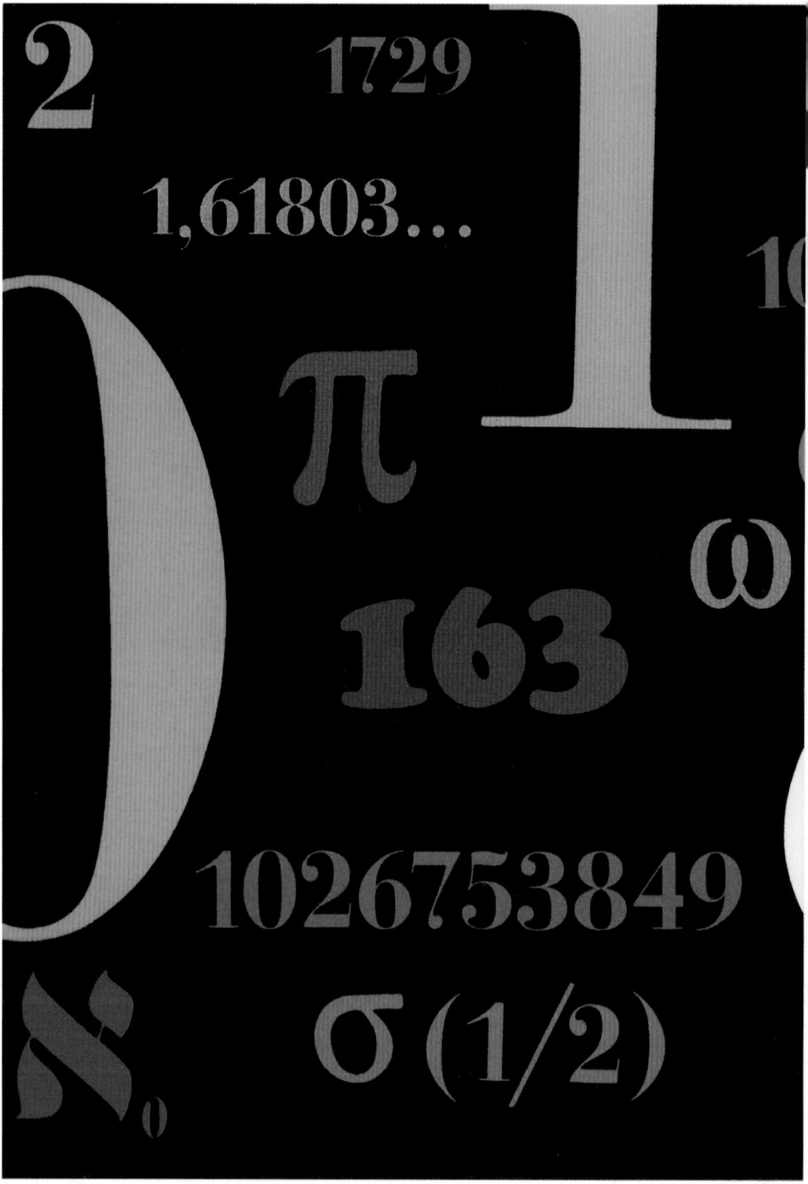

2

1729

1,61803...

1

10

π

ω

163

1026753849

σ (1/2)

ℵ₀

Over time, numbers began to follow a single, consistent notation system, and their universe expanded. On this long, open road each advance has further transformed the idea of number, whose principal role has gradually altered from quantifying to calculating. This road is the passage from arithmetic to algebra.

CHAPTER 5

THE UNIVERSE EXPANDS

"What do you mean by 'a certain number'?"
"By 'a certain number' I mean an uncertain number...and it would be neither respectful nor, perhaps, advisable to wish to determine it precisely."
Alphonse Karr,
Le Figaro, 1873

Left: numbers as graphic images, from a French book cover.

Right: Descartes's diagram of the square root.

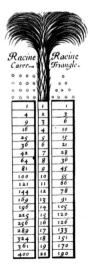

Racine Carée		Racine Triangle
1	1	1
4	2	3
9	3	6
16	4	10
25	5	15
36	6	21
49	7	28
64	8	36
81	9	45
100	10	55
121	11	66
144	12	78
169	13	91
196	14	105
225	15	120
256	16	126
289	17	133
324	18	151
361	19	170
400	20	190

Negative numbers are born from accounting

To make a measurement or work with geometric *magnitudes* one needs only positive numbers. What kind of shape or object would have a measure less than 0?

The Babylonians, Egyptians, Greeks, and Arabs did not make use of the general idea of negative numbers. The first mathematicians to do so were the Indians, who employed them for accounting purposes in the 6th and 7th centuries. In their ledgers they inscribed amounts owed as negative numbers, tallied against assets, represented by positive numbers. Eventually negatives broke away from the practical financial purposes for which they had been invented, as abstract thinkers began to explore the pure concept of negative quantity.

No tally of debts and assets can be made without a principle of equilibrium, in which assets cancel debts. Similarly, negative numbers could not exist without the prior concept of zero, the point of equilibrium.

Negative numbers in the West

A thousand years after India first employed them, negative quantities still were not in use in the West. It was the end of the 15th century before they appeared there. Why did Europe's mathematicians not adopt negative numbers in the 14th century, when they acquired the zero? Perhaps the problem was a conceptual one, difficult for the materialist traditions of European

Beginning with accounting principles, Indian calculators established what students know as the *law of signs:*

- an asset deducted from zero is a debit: $a > 0$, $0 - (+a) = -a$
- a debit deducted from 0 is an asset: $0 - (-a) = +a$
- the product or the quotient of two assets or two debits is an asset
- the product or the quotient of an asset by a debit is a debit.

Left: subtracting with tokens; below: an Indian merchant absorbed in his calculations.

debit / pay / remainder

dette / *paye* / *Reste*

philosophy to accept. Though negative entities had laws governing their use, they were not understood as real quantities—that is, as numbers, which could be considered as a possible solution to an equation. (As we shall see, equations are the central mechanism of algebra, the means by which numbers and numerical structures are defined.) Instead they were dubbed "absurd numbers"; the French philosopher René Descartes (1596–1650) called the root in an equation that turned out not to be positive a "false root." And if negative numbers did not appear in equations until

The great 7th-century Indian mathematician and astronomer Brahmagupta used colors to symbolize the different unknowns in his equations: black for the second one, then, in order, blue, yellow, white, and red.

the Renaissance, they were even slower to emerge in graphic representations: it was not until the mid-17th century that the Englishman John Wallis (1616–1703) attributed negative coordinates to the points of a curve on a graph.

Negative numbers, together with positive numbers and zero, form the set of *integers*. The set of integers is notated thus: $\mathbf{Z} = \{\ldots, -3, -2, -1, 0, 1, 2, 3, \ldots\}$; the use of the letter $\mathbf{Z}$ for this set is a convention.

Rational numbers, broken numbers

The general idea of *connectivity*—of smaller numbers that fit between two integers—was developed by the early Greek followers of Pythagoras in the 6th century BC. The Babylonians and Egyptians before them had used few of the numbers we call *fractions,* mainly those with an upper number of 1 ($\frac{1}{2}, \frac{1}{3}$, and so on), and a few other convenient ones, such as $\frac{2}{3}$. The word *fraction* comes from the Latin root *fractio,* translated from the Arab *kasra,* meaning broken: fractions are broken numbers. A fraction indicates that an integer is being divided into parts smaller than itself. The line between the upper and lower number in a fraction represents the operation of division. The lower number, called the *denominator,* names the whole; the upper number, or *numerator,* counts

It is hard to believe that all these xs and ys and the signs +, −, ×, √, which for us symbolize mathematics so completely, are fairly recent in origin. For example, the sign + did not appear in Greek or Arabic mathematics; it was invented by the Englishman Robert Recorde (1510–58) in 1557 (above: an excerpt from one of his books). It is hard to imagine the impact that the development of symbols had in the advance of algebra. Seeking to escape literal writing, scholars must have tried many kinds of notation before they arrived at the equation form we know today.

the parts. The number $\frac{2}{5}$ thus indicates 2 parts of 5: two fifths.

Integers and fractions together form the set of *rational numbers;* this set is notated as **Q**. Unlike an integer, a fraction is not necessarily a multiple of units (that is, a group of many ones). With the rational numbers, the concept of quantity becomes broader and more complex, expanding from *counting* to *measure.*

For Pythagoras, numbers ruled the universe

The followers of Pythagoras developed a true philosophy of number. They considered

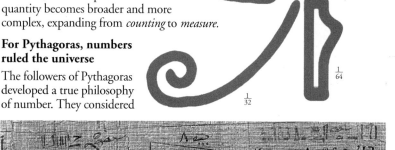

numbers not only as pure quantity but also as the constituent elements of the universe: "The principles of numbers are the elements of all entities," they declared. For them numbers were integers and the connectivities among integers, whose main role was to represent the measure of geometric magnitudes. Numbers were thus linked to magnitudes and the universe was viewed in terms of mathematical relationships. The Pythagoreans saw numbers—and indeed mathematics —in mystical terms, as a branch of philosophy or an expression of religious thought. Few mathematicians in later centuries have understood numbers in quite this elevated way; number theory has remained a discipline of profound abstraction, but has perhaps lost some of its spiritual quality.

The Pythagorean dream of perfect, universal mathemat-

Top: the numbers of the geometric progression $\frac{1}{2}, \frac{1}{4}, \frac{1}{8}, \frac{1}{16}, \frac{1}{32}, \frac{1}{64}$ are each represented by an individual hieroglyphic, which, when cleverly arranged, together compose the Eye of Horus, the Egyptian falcon-headed god.

Fragments from the Rhind Papyrus (above and below), a text on mathematics from c. 1700 BC.

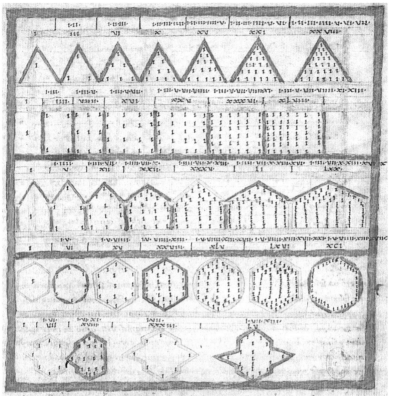

ical order was rudely broken by mathematics itself, which developed problems not susceptible to the orderly solution the Greeks desired. Disorder was found at the very heart of a key geometric figure of the ancient world: the square. This rupture was accomplished with the aid of Pythagoras's own theorem on the right-angled triangle: that in a right-angled triangle the square of the length of the *hypotenuse* (the side opposite the right angle) is equal to the sum of the squares of the other two sides. Where a and b are the right-angle sides and c is the hypotenuse, $a^2 + b^2 = c^2$. This formula, presented differently, had been known in Babylon more than a thousand years earlier, but it was the Greeks who offered a rigorous proof of it.

The Pythagoreans associated numbers with geometric figures obtained by the regular arrangement of points whose sum constituted the number represented. 1, 3, 6, 10,… were triangular numbers; 1, 4, 9, 16,… were square numbers; 1, 6, 12,… were rectangular numbers. Above: these *figurate numbers* were reinterpreted by the Roman philosopher Boethius (c. 480–524) in the 5th century AD.

The Greeks were not the first to discover the Pythagorean theorem (below: the bust of Pythagoras). They were, however, the first to offer a proof; that of Euclid is shown at left. The right triangle appears in the center; the large square whose side is the hypotenuse is below; and the two small squares whose sides form the sides of the right angle are above. The proof demonstrates that the area of the large square is equal to the sum of the areas of the two smaller squares.

In the illustration on page 85 we see that a square has two lengths: the side and the diagonal. Given the length of one, the length of the other can be determined by means of Pythagoras's theorem. Take a square whose side measures 1 and divide it into two equal right triangles whose hypotenuse is the diagonal of the square. The length of the hypotenuse must be such that its square is equal to 2, since $1^2 + 1^2 = 2$. But as the Pythagoreans proved, no rational number with a square of 2 exists. The square thus harbors within its geometry an element that is irrational in arithmetical terms. Similarly, if we divide a square with a diagonal whose length measures 1, it has a side whose square must equal $\frac{1}{2}$. And no rational number with a square of $\frac{1}{2}$ exists.

Here is a proof that the side and the diagonal of the same square do not adhere to the same rules of measure. If a number represents one, there is no number within the same system that can represent the other—the side and the diagonal are thus said to be *incommensurable*. It is impossible to know them numerically together. And yet, there they are before our eyes: a simple, perfect, and concrete geometric form. The reality of the square is not in question and it clearly transcends the capacity of numbers to express it. In other words, geometry and arithmetic are not parallel disciplines. Despite what we might intuitively expect, we cannot assume that geometric concepts will translate logically into numerical ones, or vice versa.

Rational numbers no longer express the world

For the rational Pythagoreans, this was a terrible conclusion: theory failed to express reality. The Greeks gave a name to magnitudes that eluded numericity, designating them *alagon,* the inexpressible. These are constructs for which no numerical language exists.

Even if we concede that any length with a square of 2 cannot be represented by a number, we still have to deal with that fact. In order to reconstruct the shaky edifice of mathematics the Greeks developed an internal theory concerning only irrational magnitudes. They established proportions among the magnitudes, but refused to name them *numbers.*

Almost two thousand years later mathematicians achieved a definition for these disturbing entities: all those with a square of 2 (the creature that started all the fuss) having any rational connection with the unit were designated the *irrational*

a

c

$a^2 + b^2 = c^2$

b

The irrationality of $\sqrt{2}$ is proved using the natural numbers' properties of parity: every rational number can be represented by an irreducible fraction $\frac{a}{b}$. a and b do not have a common divisor, so they cannot both be even numbers (or both would have 2 as a divisor). Suppose that there is a rational number $\frac{a}{b}$ with a square of 2: $\frac{a^2}{b^2} = 2$ thus $a^2 = 2b^2$, or a^2 is even, which means that a is as well, since only an even integer can have an even square. If a is even, we can write $a = 2c$, either replacing the equation above, $(2c)^2 = 2b^2$, and thus $4c^2 = 2b^2$, or, simplifying by two, $2c^2 = b^2$— which means that b is also even. And that is a contradiction, and thus absurd.

number "square root of 2," notated $\sqrt{2}$. (The symbol $\sqrt{}$ indicates a square root.) Thus, with a new label, such entities finally joined the universe of numbers.

The decimals: numbers with a point

A rational number has another, more flexible mode of representation besides appearing as a fraction: the decimal system. In this system, which relies directly on the principle of place value, the decimal point marks the division between unit and fraction: $\frac{1}{2}$ can be written 0.5, and 0.3333… represents $\frac{1}{3}$. It is called *decimal,* from the Latin for *ten,*

Music and mathematics are intimately related. The unfolding of music in time is regulated by the musical *measure*: a phrase of music has a standardized, measurable length. In modern musical notation, at the beginning of each stave a fraction indicates the number of beats in the measure (the numerator) and the unit value of time (the denominator). Thus, in the denominator the whole note is denoted by 1, the half note by 2, the quarter note by 4, and so on. A *time signature* of $\frac{4}{4}$, for example, indicates a time based on 4 quarter notes; one of $\frac{3}{4}$ is based on 3 quarter notes, and so on. Within each measure the organization of rhythm is variable. Thus, in the canon (top) by Johann Sebastian Bach (1685–1750) in $\frac{4}{4}$ time (notated as C), we find in the first measure 2 eighth notes, 1 half note, and 1 quarter note (4 beats), in the second 1 eighth note, 2 sixteenths, 1 half note, and 1 quarter note (4 beats), and so on. Many contemporary compositions (center) do not employ the measure, but set the rhythm as a speed (for example, 1 quarter note = 160). During the Baroque period unmeasured scores were common, leaving the interpreter broad discretion (bottom: a passage by Louis Couperin, 1626?–61).

because it uses base 10: every fraction is converted into tenths or other multiples of 10. Note that the following decimals all have the same value: .1, 0.1, 0.10, .1000; all are ways of writing *one tenth*—the 1 is always in the tenths place.

The decimal system of writing numbers was developed definitively in the 15th century by the Persian mathematician and astronomer al-Kashi (died c. 1436), director of the observatory of Samarkand, in his treatise *The Key to Arithmetic.* In the West we owe the generalized use of decimals to the Dutch mathematician Simon Stevin (1548–1620), who demonstrated them in his 1585 text *The Tenth.*

In rational numbers that are fractions, when the numerator is divided by the denominator the decimal is not always a finite number. For example, $\frac{3}{4}$ written as a decimal is 0.75, a finite number, but $\frac{5}{9}$ written as a decimal is a sequence of numbers that does not end: 0.5555555…. These sort of decimals are called *infinite decimals.* Some fractions have a truly startling property: when written as decimals a group of digits appears, endlessly repeated in the same order. For example, the fraction $\frac{1}{7}$ so written produces: 0.142857142857…, in which 142857 is repeated infinitely. (Some numbers begin their repetition after one or two digits: $\frac{21}{22}$ = 0.954545454….) These fascinating decimals are called *repeating decimals.* The decimals of many rational numbers are thus predictable.

The same is not true for *irrational numbers.* These are numbers in which numerals follow the decimal point forever, but none of the numerals after the decimal point can ever be predicted, and no pattern emerges. Irrational numbers are among the most intriguing in mathematics.

Donne
que) 9413
deſſus les
les meſme
27 ◉ 8 ①
27 $\frac{8}{10}$, $\frac{4}{10}$
raiſon les
8,75 ◉ 7 ⦵
nombres,c
enſemble
mais autan

In 1427 al-Kashi (manuscript at left) defined decimal fractions, for which he proposed a simple notation, and established the rules of calculating with decimal numbers, determining the placement of the decimal point. Above: a passage from the French edition of *The Tenth,* showing Stevin's use of fraction notation. Opposite below: Nicolas Chuquet's 1484 text *Three Parts in the Science of Numbers* was one of the seminal Renaissance works on fractions, equations, and irrational numbers.

me (par le 1 probleme de l'Arithmeti-
ui ſont (ce que demonſtrent les ſignes
bres) 941 ⓪ 3 ① 0 ② 4 ③. Ie di, que
t la ſomme requiſe. *Demonſtration.* Les
7 ③ donnez, font (par la 3e definition)
7/000, enſemble 27 $\frac{847}{1000}$, & par meſme
6 ① 7 ② 5 ③ vallent 37 $\frac{675}{1000}$, & les
② 4 ③ feront 875 $\frac{782}{1000}$, leſquels trois
ne 27 $\frac{847}{1000}$, 37 $\frac{675}{1000}$, 875 $\frac{782}{1000}$, font
le 10e probleme de l'Arith.) 941 $\frac{304}{1000}$,
t auſſi la ſomme 941 ⓪ 3 ① 0 ② 4 ③,

The most famous of them is the one called *pi*, written π, the ratio of the diameter of a circle to its circumference. (We shall revisit this number later.) Irrational numbers are difficult to represent in notation; since they are infinite but without repetition, the convention of "…" cannot be used, since it implies continuation of the numbers already given.

Real numbers express continuity

So the measures of real magnitudes are not always rational numbers. To get beyond the inability of the rationals to represent all the measures of magnitudes, the field of numbers had to be extended. In the 10th century the Persian math-

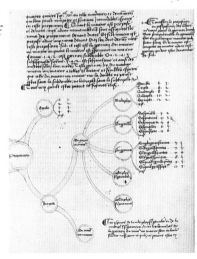

ematician and poet Omar Khayyám (1048?–1131) established a general number theory. This idea of categories or kinds of numbers is an important one. Over the centuries mathematics has continued to discover (or create) new categories of numbers and with them new ways of thinking about numbers and what they do. The identification of such types or categories is an exercise in mathematical thought.

Omar Khayyám was such a thinker. He added attributes to numbers beyond the properties of the rationals. These enhanced numbers, which comprise both rationals and irrationals, we call the set of *real numbers,* notated as **R**. But he encountered a problem when he attempted to transform the rationals into these new, more flexible entities. The usual operations of calculation—subtraction and division—had been sufficient to bring about previous expansions of the definition of numbers. For example, division had created the expansion from natural to rational numbers and subtraction had created the negatives. But there was a fundamental difference between the rational and the real numbers: the set of all real numbers forms a *continuum.* That is, real numbers may be visualized the following way: imagine a line with an origin at the point 0, running through the integers 1, 2, and so on to infinity. This may be called the *real number line.* The real numbers are *continuous* along that line. No matter how small or how "close together" two points on the line may be, there are always numbers between them—integers, fractions, numbers of all sorts. We may indicate any point *A* on the line and one and only one real number *a* will correspond to it. The sign + or – indicates the direction of the line; the number (without its

Algebra is the science of the theory of equations. Both the word and the field of study appeared for the first time in al-Khwārizmī's book *Kitab al-jabr wa al-muqa balah* (AD 825), literally: *Treatise on restoration or completion and of reduction or balancing.* Through translation *al-jabr* became *algebra.* First explored by the Greek Diophantus (3d century AD), it evolved over the course of six centuries in the Arab world, resurfacing in Europe in the 16th century among Italian mathematicians, including Niccolò Tartaglia (1499–1557) and Geronimo Cardano (1501–76).

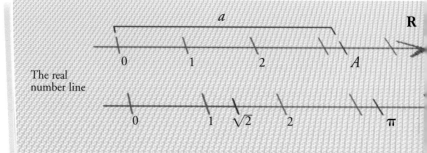

The real number line

sign) indicates the length. Thus, a real number that corresponds to the diagonal of a square with a side of 1 can give its measure on this line as $+\sqrt{2}$. Similarly, for the circumference of a circle with a diameter of 1, the measure would be the number π. Arithmetical operations alone are not sufficient for the complex kinds of calculations continuity permits. The idea of continuity and irrational numbers together led to a new mathematics of change and flux, velocity and acceleration: the calculus.

As we shall see, it was not until the end of the 19th century that the idea of the continuum was fully formulated and the set of real numbers acquired a satisfactory definition. Meanwhile, the discipline of mathematics altered to accommodate these new kinds of numbers, while the incommensurableness of geometry and arithmetic remained.

Equations: the number factory

How are new numbers produced? By using equations. A number is not identified as a random thing in itself, but as the solution of a given equation or type of equation. The natural number 4, for example, is the solution of the equation: $x - 4 = 0$. Equations are therefore a tool of number theory.

And what are the solutions to the equation $x + 4 = 0$? That depends. If we restrict the search to

Paul Colin designed this poster (above) for *The Adding Machine,* a 1927 play by Elmer Rice. Equations, so menacing here, are the poetry of mathematics: numbers in action. Today, with the advent of pocket calculators and hand-held computers, knowledge of the fundamental operations of mathematics is being lost. How many people still know how to extract a square root by hand?

R

natural numbers, we can find none. If, on the other hand, we want this equation to have a solution at all costs, we must enlarge the field of possibilities by constructing new numbers—a device mathematics permits. Starting with **N** (the set of natural numbers), we invent a new mathematical entity, –4, bound to the natural number 4 by the relation $4 + (-4) = 0$. The negative integers are thus defined as the set of solutions of all the equations of the type $x + n = 0$, where n is a natural number.

Taking the same approach, the irrational number $\sqrt{2}$ is seen as one of the two solutions to the equation $x^2 - 2 = 0$. In effect, $(\sqrt{2})^2 - 2 = 0$. The other solution is $-\sqrt{2}$, since $(-\sqrt{2})^2$ is also equal to 2.

The impossible negative square

Going further, we may ask: what are the solutions of the equation $x^2 + 1 = 0$? If there is a solution it must be a number with a square equal to –1. Because a negative number times a negative number is a positive, it follows that the square of all real numbers is positive. The equation $x^2 + 1 = 0$ thus will not permit any solutions in **R**, the set of real numbers. If we want to give it one, we must invent new numbers

The contemporary artist Jean-Louis Brau painted *Quite Square* (left). Artists through the ages have been fascinated by the imagery and symbolism of numbers.

with a negative whose square is a negative number. What kind of reality can such entities have?

In mathematics it is always possible to define new mathematical entities, on one condition: their existence must not jeopardize that of preexisting or already-established entities, otherwise they introduce a fatal logical contradiction into the structure of mathematics, destroying its entire edifice.

With his right hand the monk Luca Pacioli (1445?–1514?), author of the *Summa de arithmetica* (1494), does geometry, with his left, arithmetic. Leonardo da Vinci was his student.

$$\Phi = \sqrt{1+\sqrt{1+\sqrt{1\dots}}}$$

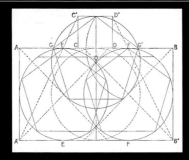

$$\Phi = 1 + \cfrac{1}{1 + \cfrac{1}{1 + \cfrac{1}{1 + \ldots}}}$$

Harmony is expressed in numbers. Whether in pictorial or architectural space, or in the realm of music, many have tried to express harmony in the language of number. In Classical thought the beautiful was said to be lodged in a wonderful number called the golden number, represented by the Greek letter Φ, $\Phi = \frac{1+\sqrt{5}}{2}$, one of two roots of the equation $x^2 - x - 1 = 0$, ordinarily used in its decimal value of 1.618. The golden number turns up everywhere, from the architecture of the Egyptian pyramids and Greek temples to compositions by Raphael, Leonardo da Vinci, Poussin, Cézanne, and Le Corbusier. Left: the structure of Rogier van der Weyden's great altarpiece *The Descent from the Cross*, c. 1435, comprises golden-number geometry, linked to a Christian symbolism in which Christ's body represents the perfect harmony of parts.

A golden section (below) expresses perfect, harmonious proportion: it is the division ("section") of a line or the proportion of a geometrical figure such that the smaller dimension is to the greater as the greater is to the whole.

a : b :: b : c

To real numbers we may now add new numbers created expressly to have a negative square. We accomplish this by adjoining to the real numbers a new entity, called i (from the first letter of the word *imaginary*). This entity is defined through the usual arithmetical operations. For example, if $i + i = 2 \times i,$ and $i^2 = -1$, what is i? It is, Leibniz answered, the imaginary root of the negative unit: $i = \sqrt{-1}$. Here we

In the 16th century Geronimo Cardano wrote that negative roots would cause "mental tortures." Above: imaginary numbers as conceived by the Surrealist painter Yves Tanguy in 1954.

have the entity whose square is equal to –1. Since it is not a real number, its presence does not contradict any established law in the universe of real numbers. Mathematicians claim that this entity is a number.

Part real, part imaginary

With the advent of *i*, more new kinds of numbers were possible. The next sort of numbers to be created were *complex numbers,* forming the set **C**. These are numbers that have both a real and an imaginary part. The imaginary part is always *i*, the square root of –1, or a multiple of it. This can be put another way: given a pair of real numbers *a, b*, the complex number *z* is defined $z = a + ib$, *a* being the real part and *b* the imaginary part of *z*.

We have already seen that each positive real number *a* has its corresponding negative –*a*. Likewise, each real number *a* has a corresponding imaginary number: *ia*.

In 1545 the Italian mathematician Geronimo Cardano was the first scholar to write a negative square root: $\sqrt{-15}$. In 1777 Leonhard Euler introduced the symbol *i* to signify $\sqrt{-1}$.

The complex plane

A Norwegian, Caspar Wessel (1745–1818), in 1797, and a Swiss, Jean Robert Argand (1768–1822), in 1806, independently proposed a graphic representation of complex numbers. Just as real numbers are represented by the real number line, complex numbers can be charted on the *complex plane.* Since a complex number is a compound of two real numbers and an imaginary number, it may sensibly be represented in a diagram that is a two-dimensional

"The divine spirit reveals itself in a sublime manner in this wonder of analysis, this marvel of an ideal world, this intermediary between being and nonbeing that we name the imaginary root of the negative unit."
Gottfried Wilhelm Leibniz

The diagram below illustrates how a complex number may be represented graphically on the complex plane. Complex number *z* is represented by a corresponding point on the plane, labeled point *M*. It is a sum of real number *a* and imaginary number *ib*. Real number *a* is located on the real number axis (**R**); imaginary number *ib* is found on the other defining axis of the complex plane, that of imaginary numbers (*i***R**), at some precise distance from the unit of origin. This graphic diagram is a convenient way of seeing that a complex number may express both position and direction.

The complex number $z = a + ib$ is a vector. The idea of the vector opens up a huge field of applications for complex numbers: in physics they have long been used for calculations involving electricity. In mechanics velocity is a vector, combining rate of change of position with direction.

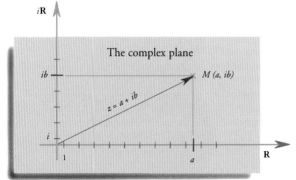

The complex plane

equivalent of the real number line. The complex plane is defined by two axes, the *real axis* (whose points are the positive and negative integers) and, perpendicular to it, the *imaginary axis* (whose points are the positive and negative imaginary numbers). Each point on this plane is a complex number, the sum of a real number and an imaginary number. No longer is there only one direction but an infinity of directions in the plane.

The complex number $z = a + ib$ is represented on the complex plane by a *vector*. The concept of the vector is of tremendous importance: it combines the idea of magnitude (a core element of mathematics) with that of direction (a core element of physics); a vector is a quantity that has both properties.

What has been lost with the development of complex numbers is the possibility of comparing two numbers. Two real numbers can always be compared: one is necessarily greater than or equal to the other. This statement no longer holds true for complex numbers. Two of these, z and z', can be such that z is neither greater, less than, nor equal to z', but simply not comparable.

Algebra redefines mathematics

By the Renaissance mathematics had achieved a level of development far beyond the practical problem-solving needs of finance, architecture, or engineering. As it freed itself of functional applications, the discipline grew in complexity and nuance and became a branch of knowledge with one foot in philosophy and one in science. To arithmetic and geometry, both grounded in associations with real quantities, spaces, shapes, and magnitudes, was now added *algebra,* a branch of mathematics based on purely abstract entities and their properties, relationships, and operations. Algebra studies numbers detached from utilitarian purposes, beautiful in themselves.

The history of numbers is a sequence of expanding definitions, from numbers used only for counting to an abstract idea of numbers, and from that to the idea of infinite numbers, thence to irrational and continuous numbers, and beyond.

N = the set of natural numbers

Z = the set of integers

Q = the set of rational numbers

R = the set of real numbers

C = the set of complex numbers

Algebra probably originated in ancient Egypt, and the Greeks had touched upon it. But true algebra was born in the Arab world in the 9th century (the word is originally Arabic: *al-jabr,* meaning *the reduction*), and came to flower in Renaissance Europe with the development of a specialized language. Instead of a calculation involving specific numbers —say, $3 + 5 = 8$—a symbolic language representing pure ideas

and processes permitted mathematicians to focus on the operations themselves: $x + y = z$. Algebra took the fundamental operations of arithmetic (addition, subtraction, multiplication, division) and the basic concepts (set theory, properties of real and complex numbers, and some others) and added rules and definitions—the "grammar" of the language—and symbols—its "vocabulary"—so that calculations could be made without ambiguity. Algebra discusses and investigates numbers in general terms, and thus reveals their nature.

The basic language of algebra is quite simple. Symbols refer to any quantity, set, or type of number, as the operation defines it; parentheses keep the order of operations clear or group certain elements together. A power is identified (as we have seen) by a smaller number attached to the main number at its upper right: 5^7 or 3^x or x^y. The times-sign has gone: multiplication is indicated by two terms or quantities placed side by side. Division is represented not by a division sign but simply by the line we first saw in fractions, dividing numerator from denominator. Over time, other symbols and terms have been added. Instead of specific or concrete quantities, we have the idea of quantity itself, expressed as *constants* (for

The young Frenchman Evariste Galois (1811–32, left) worked on problems of complex equations. By age 17 he had demonstrated a genius for higher mathematics, but could not gain the attention of the French Academy. Frustrated, he joined the French revolution of 1830. Two years later he was challenged to a duel. The night before it he wrote down his brilliant discoveries concerning equations; the next day he was killed.

Equations that express changes in quantity and rates of change use complex numbers. The solutions to such equations are also equations, rather than fixed numbers: an equation can express quantities that change, such as a rate of speed or a rate of growth, in mathematical terms, while a rational number can only express a quantity, such as speed or size at a given point. Below: an equation meets an integer.

$$x = \frac{-b + \sqrt{b^2 - 4ac}}{2a}$$

known quantities) and *variables* (for undetermined quantities) as well as the ideas of magnitude and direction.

The problem of π

A relationship between magnitudes that has always fascinated mathematicians is that represented by π. The calculators of antiquity had noticed that all circles had something in common: the diameter of any circle always has the same ratio to its circumference. Could this link be represented by a rational number? In other words, could this ratio between two lengths be known precisely, or could it only be known approximately? And, if the latter, how close an approximation could be found?

This ratio did not become a named number until the early 1700s, when it was given the designation π—the Greek letter *p*, spelled *pi*. This referred to *periphereia*, the Greek word for the circumference of a circle. Yet the problem of the ratio itself is much older. Jewish scholars around 2000 BC judged the circumference as three times the circle's diameter. In one of the oldest mathematical texts, the Rhind Papyrus (c. 1700 BC), the Egyptian scribe Ahmes tried to measure the area of a circle inscribed in a square. After conversion, the proposed value is $(\frac{16}{9})^2$, which can be written as 3.16049.... In AD 120 the Chinese mathematician Chang Hing arrived at a slightly more accurate ratio of $\frac{142}{45}$ (3.15555...). The Greek Archimedes, in the 3d century BC, offered not a value for π but a process for finding a more precise approximation. Ahmes had inscribed a circle in a square and then measured the square's area to obtain π; Archimedes inscribed a circle in a hexagon, whose perimeter he measured. He then doubled the number of

$\frac{c}{d} = \pi$

c

d

Left: if *c* is the circumference of the circle, and *d* is the diameter, π is expressed as the ratio $\frac{c}{d}$ = π. Below: a 19th-century depiction of Archimedes interrupted at his work by the troops of Marcellus, painted by Christophe-Thomas Degeorge (1786–1854). Plutarch described the Greek mathematician's death during the capture of Syracuse.

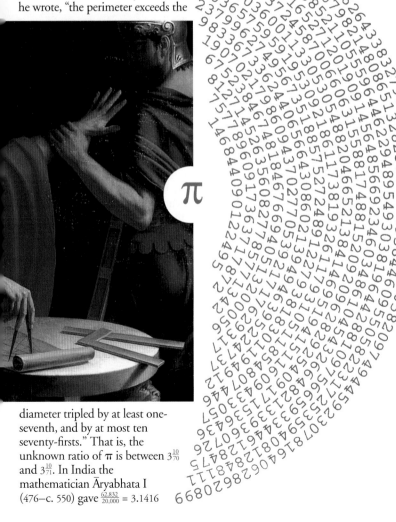

sides of this polygon and repeated his computations, continuing in this way until he was able to compute the perimeter of one with 96 sides— that is, closer than a square is in area to a circle. "For every circle," he wrote, "the perimeter exceeds the

The first few decimals of π. Note—if you have the patience—that they display no trace of periodicity.

diameter tripled by at least one-seventh, and by at most ten seventy-firsts." That is, the unknown ratio of π is between $3\frac{10}{70}$ and $3\frac{10}{71}$. In India the mathematician Āryabhata I (476–c. 550) gave $\frac{62,832}{20,000} = 3.1416$

as an approximation for π. In the 1650s John Wallis, playing on the doubling of even and odd numbers, proposed a strange fraction for π: $\frac{2}{\pi} = \frac{1 \times 3 \times 3 \times 5 \times 5 \times 7 \times 7 \times 9 \times \ldots}{2 \times 2 \times 4 \times 4 \times 6 \times 6 \times 8 \times 8 \times \ldots}$. Leibniz used only odd-numbered denominators and alternated additions and subtractions for another expression of π: $\frac{\pi}{4} = 1 - \frac{1}{3} + \frac{1}{5} - \frac{1}{7} + \frac{1}{9} - \ldots$. And recently, the first 16 million decimals of π have been calculated.

The impossibility of squaring the circle

Though it had been "easy" to prove that the ratio of the side of a square and its diagonal could not be represented by a rational number, it turned out to be difficult to prove that π was a similar kind of irrational number. The German mathematician Johann Heinrich Lambert (1728–77) established the irrationality of π in the second half of the 18th century. After this, any hope of representing π as a fraction, and thus with a periodic decimal development, proved out of the question.

A century later, in 1882, another German mathematician, Ferdinand von Lindemann (1852–1939), definitively resolved one of the oldest questions in mathematics, that of squaring the circle. The question the ancients had posed was this: using a ruler and a compass, is it possible to construct a square with the same area as a given circle? The countless attempts to do so had all ended in failure. Why? Because no exact measurement can be taken of the circle's circumference, since its length is a multiple of the irrational number $\sqrt{\pi}$, an infinite and therefore inexact number.

The other side of the mirror

Pressed by necessity, mathematicians have dared to write and do things not permitted. To do so they have passed beyond the mathematics universe of their time. One may go through the looking-glass into the world of negative, irrational, and imaginary numbers—always provided one

A 1939 costume sketch for the character Arithmetic from Maurice Ravel's dream-opera *L'Enfant et les sortilèges*.

The expression "squaring the circle" has come to signify any fruitless attempt to resolve a problem.

HISTOIRE
DES RECHERCHES
SUR LA
QUADRATURE
DU CERCLE;

Ouvrage propre à inftruire des découver-
tes réelles faites fur ce problème célé-
bre, & à fervir de préfervatif contre
de nouveaux efforts pour le réfoudre :

*Avec une Addition concernant les problémes
de la duplication du cube & de la trifec-
tion de l'angle.*

Left: the title page of Jean-Etienne Montucla's *History of Inquiries on Squaring the Circle,* a work exploring past and contemporary attempts to solve "this celebrated problem." Montucla (1725–99) was one of the first modern historians of mathematics.

Three great problems of antiquity were to square a circle, to trisect an angle (divide an angle into three equal parts), and to duplicate a cube (to find a cube whose volume is twice that of a given cube). Below: Archimedes's attempt to square the circle.

can return. But pure writing is no more possible in mathematics than in poetry and literature. To write the "impossible" is simply to pose the question of its existence in another way. In mathematics, this means elaborating a theory in which that writing represents a definite object. Irrational, impossible, absurd, broken, imaginary, real, complex, transcendental, transfinite, surreal—all these qualifiers say much about the nature of our relationship with the entity that we call *number.*

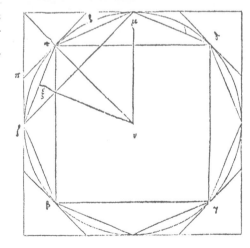

O ne, zero, infinity—these three concepts are the foundation of the universe of numbers. Zero is unique, infinity multiple. In mathematics there is only one way to be nothing, but there are infinite ways to be infinite. Zero was invented in India in the 5th century; infinite numbers were not mathematically defined until the end of the 19th. Georg Cantor (1845–1918), the creator of these infinite numbers, declared, "Nothing…can hold us back from the creative process of making new numbers."

CHAPTER 6
FROM ZERO TO INFINITY

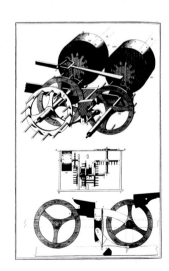

Z ero, one, many: three concepts of number upon which great edifices are built. Left: zero painted by Jasper Johns (b. 1930).

R ight: plans for Blaise Pascal's 1642 Calculating Machine, from Denis Diderot's *Encyclopedia,* volume 2, 1762–77. The invention of such machines, forerunners of computers, gave mathematicians the opportunity to make computations of immense size and complexity, but it was human imagination that brought about the revolution in ideas that is the soul of mathematics. A computer can count forever, but only a human mind can envision infinity.

1: the raw material of multiplicity

Most number systems have no zero and only one has any kind of infinite number, but no system could do without the number 1. Without it, indeed, number itself could not exist. *One* represents being, existence. A thing exists if it is, and if it is, it is one. Euclid begins his arithmetical definitions, in Book VII of the *Elements,* with the definition of the unit: "A unit is that by virtue of which each of the things that exist is called one."

From the unit comes plurality. When it has no limit Euclid calls plurality, or multiplicity, *plethos,* a general term for the many, as opposed to the object of particular knowledge. But when this "manyness" has a defined limit it is called *arithmos,* or number, whose study is called arithmetic. "A number is a multitude composed of units," Euclid states. The number 1, relentlessly reiterated, designates other numbers. As we have seen, these are used as base numbers, establishing the concept of *orders* of units, which makes the complex operations of enumeration easier.

0: a three-stage history

Zero is conceptually different from other numbers, for it is not tied to objects. "Ever since we first sought number in the object, the series of numbers has begun with 1," observed the Swiss psychologist Jean Piaget (1896–1980). "Making zero the first of the numbers means no longer abstracting them from the object." Piaget noticed that when the concept of zero is introduced number becomes an idea in itself, detached from concrete reality. It is zero that brings about this revolution in thinking.

Zero made a long journey to reach this conceptual point. To become the number we know today, it passed through three stages: symbol of notation, digit, and finally actual number. In the first

"Whatever ideas there may be of the number one in individual souls, they are still to be as carefully distinguished from the number one, as the ideas of the moon are to be distinguished from the moon itself."
Gottlob Frege,
The Basic Laws of Arithmetic, 1893

For the Greeks, *one* was not a number but the very idea of presence, of existence itself. Number began after the first numeral, with the idea of "more than one."

A 1 drawn by the Art Deco illustrator Erté.

instance zero was simply functional, a useful symbol that was not itself a number. Placed at the end of a number it multiplied it (by ten, if the calculation was made in base

Justo Gonzalès Bravo, *Untitled.*

10). For example, the number 123 with a 0 added produces the number $123 \times 10 = 1,230$.

Zero as a digit was still a mere tool. When figures began to be set in columns, separated by vertical lines and using the principle of place value, a zero became necessary. A figure was represented by one of the nine digits placed in each of the different columns, signifying the quantity of units, tens, hundreds, and so on. In a case where no quantity of a particular power of the base occurred, the corresponding column remained unoccupied, empty. It became convenient to represent this absence with a graphic symbol, so that every column, empty or full, was occupied by a sign. The zero served as a separator, making it possible to remove the grid of lines separating the columns. When the grid apparatus was abandoned, the symbol representing absence remained and became a digit like the other nine.

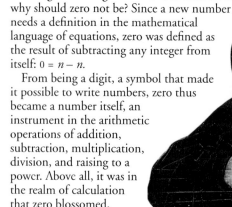

Zero as a number is a different animal altogether, with greater purposes and deeper meanings. If the digits from one to nine are numbers, why should zero not be? Since a new number needs a definition in the mathematical language of equations, zero was defined as the result of subtracting any integer from itself: $0 = n - n$.

From being a digit, a symbol that made it possible to write numbers, zero thus became a number itself, an instrument in the arithmetic operations of addition, subtraction, multiplication, division, and raising to a power. Above all, it was in the realm of calculation that zero blossomed,

Formed by a leisurely stroke, a plain circle—emblem of unity and simplicity—becomes something to reckon with. Zero has had a bad reputation among school-children. Students find it the most difficult number to understand. Up to the age of six and a half, 25 percent of children write: $0 + 0 + 0 = 3$; up to the age of eight and a half, 50 percent write: $0 \times 4 = 4$. Worse, a zero, vigorously inscribed by a baleful teacher on a poor test or paper, is the dreaded failing grade. Neverthe-less, zero is the source of many delights. Left: an anthropomorphic zero from 19th-century Italy. Below: a pharmacist struggles with calculations; opposite: the artist Miguel Chevalier creates a postmodern zero, *Binary 3D,* 1989.

creating fascinatingly distinct effects in different operations. For example, while it is completely powerless in addition— $n + 0 = n$—it is all-powerful in multiplication: $n \times 0 = 0$. In raising to a power it has an interesting impact: if a is different from 0, $a^0 = 1$. But dividing by 0 is the supreme prohibition: $\frac{n}{0}$ is an undefinable number.

The history of zero

The first zero is Babylonian and dates to before the 3d century BC. Babylonian scribes notated units as vertical or horizontal wedge shapes, or chevrons, and used a tilted double chevron as a sign indicating separation in the notation of numbers—a true digit for zero. In astronomy and for some other uses this same symbol also functioned as a tool of calculations. It was found in the first or last position among notated numbers, especially in the notation of sexagesimal fractions. At no time, however, was zero used as a number itself.

During the 1st millennium AD Maya astronomers developed an effective positional numeration system with a base of 20, in which numbers were represented by groups of dots and horizontal bars, arranged vertically. A horizontal oval glyph, representing a snail shell (or,

Dividing by zero doesn't work: if $\frac{a}{0} = b$ (where neither a nor $b = 0$), then $b \times 0 = a$. But *any* number $\times 0 = 0$, so we have a contradiction.

according to some, that of a clam), is used as a separator sign, making possible the writing of unambiguous numbers. Although the Maya zero did not function as a tool of calculation, much less bear the complex attributes of number, it remains a remarkable invention.

From emptiness to nothing

To India goes the credit for inventing the complete zero, with all three functions of place notation, designation of quantity, and number. It first appears in manuscripts of the 5th century AD. The first Indian notation for zero was a small circle, *sunya,* the Sanskrit name of the mark for emptiness. Translated into Arabic this word became *sifr* (close to the English word *cipher*); in Latin this became *zephirum* and then *zephiro,* zero.

Emptiness is a philosophical idea, but it is also a spatial category, although one that is admittedly difficult to locate. The creation of the digit zero, marking an empty place in the calculation column with a symbol, was a great innovation in mathematical thinking. It showed an absence by a presence, signifying a change in perspective from negation to affirmation.

The concept *nothing* belongs to the category of existence. The creation of the number zero merged the two categories of nothingness—spatial emptiness and philosophical nonexistence—and effected a radical transformation in the status of number. To "there is not anything" was added "there is nothing." This was no mere change in syntax, but an advance from the zero of logic to the zero of arithmetic, which has (or is) a value. The passage from "there is not" to "there is zero," from zero as the empty position to zero as the null quantity, constitutes a major step in the history of thought. We ask: how many? And we have an answer: zero!

"The Indian zero stood for emptiness or absence, but also space, the firmament, the celestial vault, the atmosphere and ether, as well as nothing, the quantity not to be taken into account, the insignificant element.**"**

Georges Ifrah,
From One to Zero: A Universal History of Numbers, 1985

Zero has been described as "this nothing that does everything!" This strange digit did not come into the world without causing some problems. For example: is 0 odd or even? An integer is defined as even if the result when halved is an integer, but this rule is useless when applied to zero. Other approaches must be sought to determine the answer. Let us turn to a concept called the conservation of parity, which states that the sum of two even numbers is even. If we decide to extend this rule to zero, then it must be considered even.

In a 1485 arithmetic text intended for merchants we read, "In digits there are but ten figures, of which nine have value and the tenth no value, but it sets off the other numbers and it is called zero or digit." Left: an immense sundial from Jaipur, India, takes the form of a zero.

Below: various forms of the Maya zero: glyphs representing what are believed to be snail or clam shells.

From the unlimited to the infinite

Through the centuries the Greeks cultivated the concept of *apeiron,* that which is boundless or infinite. They used this idea in their explorations of time and space, of the creation and disintegration of

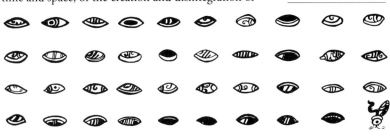

things, and in the study of numbers themselves. For them, as for us, the infinite was a concept both mathematical and spiritual; its quality of limitlessness and implied vastness has something in common with ideas of eternity, of immortality, and of deity. Time has neither beginning nor end; space, equally resistant to definition, is the abode of lines and planes whose magnitudes may be divided or extended without limit. As to the march of numbers, who can interrupt their succession?

The philosopher Aristotle (384–322 BC) was interested in the idea of infinity as an aspect of his inquiry into the nature of the physical world. Because it was both an important concept and a vague one he sought to formulate some defining terms for it. His definitions grapple with this highly abstract idea in terms of logic. First, he said, the infinite exists in nature and can be identified only in terms of quantity. Second, if it exists, it must be defined; and third, since the infinite cannot be understood as a totality it cannot exist in actuality, but only as a potential thing. With a lack of foresight rare for him, he concluded that mathematicians had no use for the concept of infinity.

A boundless number of infinities

On the question of the infinite the Classical world had other thinkers less timid than Aristotle, among them Anaxagoras of Clazomenae (c. 500–c. 428 BC) and Epicurus (341–270 BC). Democritus of Abdera (c. 460–370 BC), famous for his prescient concept of matter as composed of infinitely small and varied atoms in infinite numbers, asserted that the whole is unlimited by the number of these bodies. In support of this proposition the Roman Lucretius (c. 100–c. 55 BC) considered and attempted to refute the reverse hypothesis: that the world is finite. He proposed that we imagine an archer who stands at the extreme edge of the world and shoots an arrow. As he does so, Lucretius asks us, "Will this missile, released with great force, follow the course on which it was aimed? Or do you think something will block its way?" Before we can reply he cautions us, "Wherever you may place the ultimate limit of the world, I will ask you:

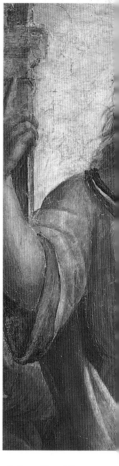

Above: Plato debates Aristotle. Opposite: an early diagram of the solar system according to Eudoxus and Aristotle. Even without modern mathematics, the ancients imagined the infinite universe to have a structure.

'What will happen to the arrow?' "
The point is that the boundary of
the real cannot be fixed, and this
conclusion is confirmed by the fact
that attempts to negate it can be
endlessly defeated.

Not content to propose the
existence of an infinite world, Lucretius
then imagines the existence of an infinite

number of such worlds. Intriguing speculations of this kind continue to provoke physicists, astronomers, theologians, and mathematicians today.

The finite in unlimited expansion

Indeed, for two millennia Aristotle's notion of an infinity that is purely potential—an infinity eternally expanding whose limit is never reached—was the most widely accepted idea on the subject in the West. This is Aristotle's elegant formula: "The infinite turns out to be the contrary of what it is said to be. It is not what has nothing outside it that is infinite, but what always has something outside it." Space is without limit, or has a supposed limit that can be "pushed back" forever. It is therefore possible to approach the infinite but impossible to attain it. In mathematical terms, for every n there will always be an $n + 1$. The infinite may thus be seen as the finite in unlimited expansion. Like other numbers, the potential infinite has been given a symbol to represent it, ∞, a limit we approach but can never attain.

Aristotle's proposition rejecting the existence of an actual infinite became a staple of exercises in logic. The argument refuting an actual infinite posited the obvious: since infinity is an all-embracing incompleteness, its existence would require a logical contradiction. This actual infinite, described in terms of number, would find itself simultaneously even and odd, divisible and indivisible, which is illogical.

Nevertheless, the idea of an actual infinite continued to attract interest among thinkers. At a certain point in the history of numbers the philosophical and religious meanings of infinity began to diverge from the mathematical. In humanistic terms the infinite remained

What are the limits of the universe? What is its form? Above: M. C. Escher's perplexing visual conundrum *Circle Limit IV (Heaven and Hell),* 1960, proposes a universe of paradoxes.

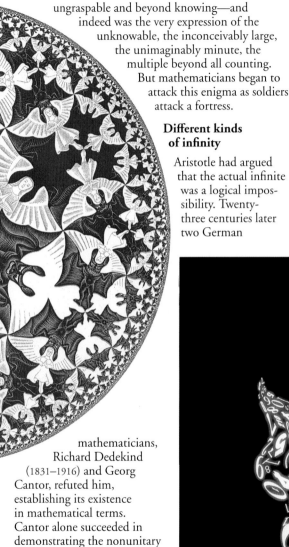

ungraspable and beyond knowing—and indeed was the very expression of the unknowable, the inconceivably large, the unimaginably minute, the multiple beyond all counting.

But mathematicians began to attack this enigma as soldiers attack a fortress.

Different kinds of infinity

Aristotle had argued that the actual infinite was a logical impossibility. Twenty-three centuries later two German mathematicians, Richard Dedekind (1831–1916) and Georg Cantor, refuted him, establishing its existence in mathematical terms. Cantor alone succeeded in demonstrating the nonunitary nature of this infinity.

Aristotle's central assertion, from which flows his

Georg Cantor wrote: "I experience true pleasure in conceiving infinity as I have, and I throw myself into it.… And when I come back down toward finiteness, I see with equal clarity and beauty the two concepts [of ordinal and cardinal numbers] once more becoming one and converging in the concept of finite integer." For many mathematicians the encounter with pure number is an exhilarating and spiritual experience. Below: digits form an abstract composition.

conception of the infinite, is a statement of the fundamental relationship between the finite and the infinite: "The whole is greater than the part." This simple declaration has the ring of a self-evident truth. A whole is a whole because it contains its parts. A part seems to be, by definition, smaller than the whole to which it belongs. A part that contests the unifying superiority of the whole is illogical; it disqualifies itself and loses its property as part. Whether functioning as an explicit or implicit axiom, this assertion firmly closes the door to infinite number, as Aristotle conceives it.

Cantor and Dedekind looked at the problem from an entirely different point of view. In order to define the infinite, they used an operation called a *one-to-one correspondence.* This disarmingly simple exercise became, in their hands, a formidable weapon. Imagine two separate sets of things, each of some unknown quantity. Imagine making pairs of things pulled one from each set. It is possible to understand that *if the sets each have an infinite quantity* the process of pairing can go on forever. If one set is not infinite, then the process ceases. Those sets that become exhausted at the same time have something in common: we may say of them that they each have a certain quantity of elements, and that the two quantities are equivalent.

Cantor and Dedekind made one-to-one correspondence the foundation of their exploration of the concept of infinity. They argued that two sets placed in such a correspondence may be said to have the same number of

"I have not ignored the fact that by this undertaking I find myself in opposition, to a certain degree, to the widely accepted conceptions concerning the mathematical infinite and to the points of view that have often been adopted regarding the essence of numerical magnitude."

Georg Cantor,
*Foundations of a
General Theory of
Manifolds,* 1883

elements, even if both sets are infinite and therefore their quantity is unending: two sets in one-to-one correspondence are equivalent. Establishing this was the first step toward a new view of the mathematical infinite.

Infinity: the part is equal to the whole

In the 1870s and 1880s Cantor and Dedekind brought about a dramatic reversal of Aristotelian logic. What had always been understood as an impossibility—that the part is equal to the whole—was for them a basic axiom, indeed the very principle that defined infinity. Their key assertion was, on the face of it, illogical: that a set that is equivalent to one of its own parts is infinite.

How can a set be equivalent to one of its own parts? What is an infinite set and how can we see that it exists? We have already encountered many such sets. The set of positive integers, $\mathbf{N}$, is infinite: one can count numbers from 1 onward forever, as we have seen. The set of even integers $\mathbf{P}$ is a *subset,* or part, of the set of positive integers. We can construct a one-to-one correspondence between $\mathbf{N}$ and $\mathbf{P}$. Here it is: for every integer of $\mathbf{N}$ let us place in correspondence its double, which is even, thus belonging to $\mathbf{P}$. Inversely, for every even integer of $\mathbf{P}$ let us place in correspondence its half, which is a positive integer, thus belonging to $\mathbf{N}$. Furthermore, the set of positive integers $\mathbf{N}$ is not greater than the set of even integers $\mathbf{P}$, even though $\mathbf{P}$ is a part of $\mathbf{N}$. Both are infinite. This sort of infinite set is called *denumerable,* or sometimes *countable:* a set that goes on forever, but can still be counted, if one is willing to count forever. The set of all real numbers, in contrast, cannot be counted because between any two real numbers there is always another real number, so that one cannot even begin to count: the set is both infinite and *not* denumerable. Here we have an actual infinity—a kind of infinity that demonstrably exists.

R ichard Dedekind (shown at left), a rigorous but open-minded thinker, was one of the rare scientists in the 1880s who did not shrink from an arithmetical treatment of infinity. He maintained a close correspondence with his colleague Georg Cantor (opposite) for twenty-seven years, from 1872 to 1899. This continuous dialogue in letters is a remarkable set of documents, almost unique in the literature of mathematics, in which the development of ideas between two of the discipline's great minds may be traced and studied. Each showed the other the evolution of his thinking and received perceptive criticisms, still illuminating to students today.

O verleaf: in his *Visible Poems* the Surrealist artist Max Ernst (1891–1976) depicts the notion of pairing. On the left-hand page each left eye stares at a right eye, forming a pair; on the right-hand page each right hand grips a left hand with fraternal warmth, making a pair.

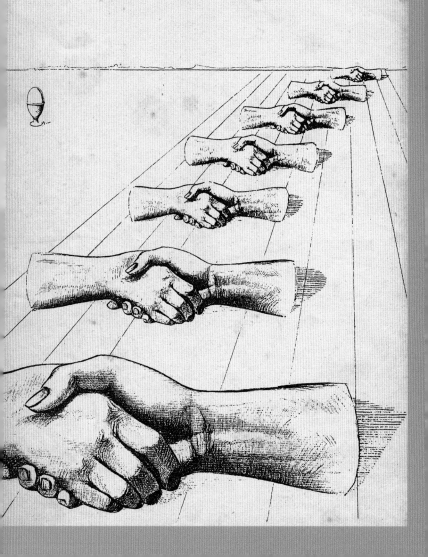

The mathematics that arose from Cantor and Dedekind's propositions demolished several age-old cherished assumptions. For example, to the astonishment of all and in contradiction to our intuitive sense of the concept of number, it turns out that there are *not* more fractions than there are positive integers. Cantor established, in effect, that the set of rational numbers **Q** is equivalent to the set **N**.

The continuum

In identifying a countable infinity, and in using the idea of one-to-one correspondence to do it, Cantor formulated the notion that infinite sets could have different *orders* or *degrees* of infinity—that one could distinguish a hierarchy of infinities. The set of real numbers, **R**, demon-strates this. As we saw earlier, real numbers form a continuum, an imaginary line running to infinity in both directions, along which numbers lie with infinite density. For every two points on the continuum there is always an infinite number of points between the two. The continuum is in this sense *unlimited*. We observed the continuum when we first noticed real numbers, but its uncountable nature can be seen only when contrasted with a countable infinity.

It is impossible to construct a one-to-one correspondence between the sets **N** and **R**, for there are "infinitely" more points on a line—and therefore more real numbers—than there are positive integers, even though both sets are infinite (and even though the term "more" seems a little misleading when used in reference to any sort of infinity). This gives us two different kinds of infinite sets, and the unlimited, uncountable set of real numbers has a higher degree or order of infiniteness than the set of integers, by virtue of its infinite density, its

Above: a photograph of a quasar. How large is infinity? How many stars are there in the universe? For every imaginable limit to space there remains a space beyond the limit. Below: Cantor suggested using *aleph*, the first letter of the Hebrew alphabet, as the symbol for his new kind of infinite number, which he called *transfinite*.

continuity. For there are no "more" points in a line segment than there are in the entire line of which it is a part! That is the extraordinary nature of infinity. Cantor speaks of the "marvelous power of the continuum."

Are there other kinds or orders of infinity besides the countable and the continuum? Cantor explored the idea of orders of infinities in great depth, using the concept of infinite sets and subsets, and came to the astounding conclusion that there is an infinity of infinities.

After this research, the finite came to be defined by the infinite, instead of the other way round. "That which…is not infinite is finite," Cantor said; that which cannot be placed in one-to-one correspondence with one of its own parts is finite. The finite, defined by what it is not, and is not able to do, comes to be seen as one of many possible concepts of number, rather than the central one. The finite was a part that, until Cantor and Dedekind constructed their new mathematics of the infinite, had been taken for the whole.

This architecture of infinities is extraordinarily spacious and grandiose—so much so that it both intoxicates and dizzies us. It was called "the most astonishing product of mathematical thought, one of the most beautiful realizations of human activity" by the German mathematician David Hilbert (1862–1943).

"Treat the laws and the relationships of integers like those of the celestial bodies," Cantor wrote. His original thinking on the nature of infinity was a radical departure from orthodoxy, disturbing to both traditional theologians and conventional mathematicians. Below: Cantor perches on the infinite (the *aleph* symbol), while balancing God in his cloud and his own arch-critic, the mathematician

Leopold Kronecker. Both, the cartoonist implies, are his implacable enemies. But mathematics shares some elements with theology. "My theory is solid," Cantor stated. "I have drawn [its] principles from the first cause of all created things."

The universe of numbers continues to expand. In our complex world numbers hold a preeminent place, colonizing every aspect of our lives. We have come to see number as the new god of modernity, responsible for counting and accounting for all things material. If the work of numbers is reduced to the menial tasks of quantification, then numericity represents a diminishment of our culture; surely this is not what we desire. Number is too beautiful a human invention to be reduced to such purposes.

CHAPTER 7
THE IMPOSSIBLE DEFINITION

In antiquity, when the system of commercial barter and exchange was replaced with the invention of money, numbers acquired a pivotal role in society. Stamped in metal, printed on bills, or written by hand on letters of credit and checks, numbers express value. Left: *Binary Spiral*, by Miguel Chevalier. Right: the tyranny of numbers: a man struggles with his tax returns in a drawing by Siné, and the numbers win.

Divine and human numbers

The German mathematician Leopold Kronecker (1823–91) once said, "God made the integers, all else is the work of man." First causes, this comment suggests, are divine, as number itself is divine, while the complexities, minutiae, and refinements of mathematics are a human creation. For Kronecker's contemporary Dedekind, however, the integers too were the "free creations of the human mind." "What are the numbers and what is their purpose?" he asked in 1888. For him, as for many modern mathematicians and theorists, mathematics

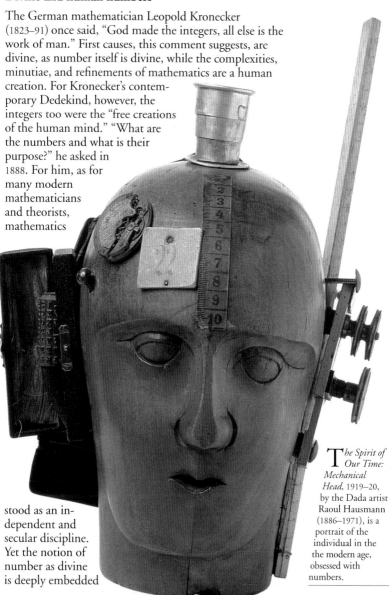

stood as an independent and secular discipline. Yet the notion of number as divine is deeply embedded

The Spirit of Our Time: Mechanical Head, 1919–20, by the Dada artist Raoul Hausmann (1886–1971), is a portrait of the individual in the the modern age, obsessed with numbers.

in many religions, from the numerological symbolism in the medieval Jewish text called the Kabbalah to the ninety-nine names of God in Islam to the doctrine of the Trinity in Christianity. Number, with its universal language and internationally understood concepts, has the power to bridge all cultures. Its abstraction, its detachment from material reality, and its association with such great philosophical questions as the idea of perfect form, the definitions of being and nothingness, and the nature of infinity place it in close relationship to religious or spiritual thought.

The revolution in number theory and other branches of higher mathematics that took place in the 19th century was part of a long process of development in modern scientific method that included great changes in physics, the natural sciences, and natural history and led to fierce debates about secular and religious issues. The dramatic growth in theoretical mathematics took place within this context.

By the 20th century myriad theories had been elaborated by mathematicians, logicians, psychologists, and philosophers in response to these fundamental questions. Many thinkers have continued to try to

Quantification and measurement were the main tools of the sciences in the Renaissance, as Pieter Brueghel suggests in his engraving *Temperance* (above), lampooning the scientist's fondness for measuring all things within reach. This passion for measurement, beginning in the 16th century, posed some delicate philosophical problems: the measure of a thing is not the thing itself but only a gauge. Measurement contributes to knowledge, but only in conjunction with other kinds of information. An IQ is not a person's intelligence, but only an indication of some aspects of it. Too often quantity is confused with meaning.

construct systematic foundations for the universe of numbers. All such systems, however, remain open to debate and criticism, and none has definitively succeeded.

The impossible definition

Numbers have been in use for six millennia without interruption, yet we remain unprepared to define just what a number is. In its disarming simplicity number rises above every attempt to reduce it to a definition, to account fully for its many attributes, to describe its essence. Number is resistant to definition because it is an irreducible element of thought, a basic constituent.

In the 5th century BC the Greek intellectual Philolaus stated: "Without number we can understand nothing and know nothing." Twenty-five centuries later, the modern French philosopher Alain Badiou offers an opposing argument: "What arises from an event in perfect truth can never be counted."

The primacy of numbers

To count, tally, calculate, and measure; to compute, reckon, encode, decode, classify, and quantify; to enumerate, estimate, and tabulate; to arrange in a sequence or hierarchy or order: number is essential to our management and understanding of life. Number is the natural language of the sciences, integral to a thousand branches of study: meteorology, probability, statistics, demographics, accounting, strategy, economics, psychology, ballistics, engineering…the list is endless. But equally, number is crucial in the arts, in aesthetics and design, in our sense of balance and beauty, harmony, unity, and symmetry.

Yet in the modern world number has acquired an ominous presence. The more number is used to control information, to keep track of individuals, and to quantify life in general, the more the *quality* of life risks being diminished. The search for truth that is the highest aim of mathematics becomes dangerously identified with the reductive calculation of figures. We are overwhelmed by numbers: taxes, indices, percentages, differences and averages, stock listings and market prices, polls and statistics, bills, coefficients, calibers, frequencies,

Numbers in the real world: a page from *The Wall Street Journal* of 8 September 1997, showing stock-market prices for mutual funds.

Colo Bonds 9.54	...	+ 5.8
Columbia Funds:		
Balance 22.98	− 0.03	+ 15.2
ComStk 23.70	+ 0.01	+ 23.8
Fixed 13.16	− 0.02	+ 5.1
Govt 8.25	...	+ 3.6
Grth 38.12	+ 0.05	+ 24.0
HiYld 10.24	...	+ 8.8
IntlStk 15.93	+ 0.10	+ 14.9
Muni 12.33	...	+ 4.9
ReEEq 18.20	+ 0.06	+ 14.8
SmlCap 16.99	+ 0.12	+ 30.8
Specl 22.91	+ 0.13	+ 15.4
Commerce Funds:		
Balanced 26.83	...	+ 16.8
Bond 19.05	− 0.02	+ 4.6
Growth 35.63	− 0.03	+ 26.5
GroInc 21.93	− 0.01	NS
IntlEq 23.27	+ 0.30	+ 7.2
MidCap 33.14	+ 0.14	+ 15.7
STGovt 18.31	...	+ 3.8
Common Sense 1:		
Govt 1 10.36	− 0.01	+ 4.7
GrInc 1 20.59	...	+ 23.3
Grwth 1 21.18	...	+ 24.2
MunBd 1 14.05	...	+ 4.7
Common Sense A:		
EmGr A p 22.53	+ 0.15	+ 22.6
Govt A p 10.36	− 0.01	+ 4.6
GrInc A p 20.58	− 0.01	+ 23.1
Grwth A p 21.14	...	+ 23.9
InEq A p 18.36	+ 0.30	+ 9.0
Common Sense B:		
EmGr B † 22.10	+ 0.14	+ 22.0
Govt B † 10.36	− 0.01	+ 4.1
GrInc B † 20.54	− 0.01	+ 22.4
Grwth B p 21.02	− 0.01	+ 23.3
InEq B † 18.04	+ 0.29	+ 8.3
Compass Fund Instl:		
BalancedI 18.06	− 0.03	+ 18.4
CoreBdI 9.72	− 0.01	+ 4.8
LgCpGrI 18.79	...	+ 27.6

amounts, dividends, the size of this or that, the value in round figures of thus and such. We have made number responsible for expressing all things material. Is it too farfetched to speak of a dictatorship of number?

Two philosophers shall have the last word: "Grown-ups love figures," says the child hero of Antoine de Saint-Exupéry's *The Little Prince* (1943). "When you tell them that you have made a new friend, they never ask you any questions about essential matters. They never say to you, 'What does his voice sound like? What games does he love best? Does he collect butterflies?' Instead, they demand, 'How old is he? How many brothers has he? How much does he weigh? How much money does his father make?' Only from these figures do they think they have learned anything about him." Twenty-five centuries earlier, Plato had warned in *The Republic:* "It occurs to me, now that the study of numbers has been mentioned, that there is something fine in it, and that it is useful for our purpose in many ways, provided it is pursued for the sake of knowledge and not for huckstering."

Insurance number, bank-account number, telephone number, driver's-license number, social-security number… every individual, assigned an identification number of some sort, may be traced in public and private computer data banks. Number establishes order: it both gives us easy access to services and information and constitutes a genuine threat to our privacy and civil liberties. The fear of losing one's identity beneath a number is a perennial source of modern anxiety —and not without reason. A generation ago Nazi concentration-camp victims were forcibly tattooed with permanent identity numbers in an act of terrible and dehumanizing degradation. We should all retain the

memory of this indelible mark as a sign of how powerful and dangerous numbers can be.

Overleaf: the marvelous universe of numbers, as interpreted by the artist Justo Gonzalès Bravo.

DOCUMENTS

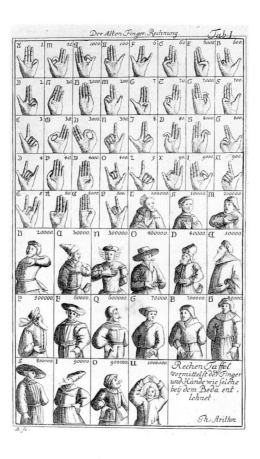

Counting

How large is large? How small is tiny? Counting can tell us the quantity of a group of things, but algebra can tell us about the quantities of things too vast to count.

Previous page: an old German engraving demonstrates a system for counting with hand gestures; above: two ancient Roman counters in which numerals are indicated with finger gestures.

Really big numbers

In the late 3d or early 2d century BC the Greek Archimedes developed a system of notation that allowed him to reach a colossal number, which today we would notate as 1 followed by 80 million billion zeros. In a text called The Sand-Reckoner *he named this number* myriakismyriostas periodou myriakismyrioston arithmon myriai myriades, *that is, a myriad myriad units of a myriad-of-myriadth order of the myriad of myriad-of-myriadth period. He set out to show that the grains of sand in the universe are fewer than this number and that therefore his system was equipped to enumerate the universe.*

Positional numeration is the mechanism that allows us to write numbers as big as we like. Is there a number so big it cannot be reduced to direct notation? No, but there are numbers that—even in modern algebraic notation—are inconveniently long.

• A *googol* is the number 10^{100}, that is, 1 followed by 100 zeros. The word was coined in 1955 by Milton Sirotta, the nine-year-old nephew of the American mathematician Edward Kasner. A *googolplex* is the number one followed by a googol of zeros!

• 1,000,000,000 (1 followed by nine zeros): the name in the United States for this number is a *billion*. In Britain, where American terms for large numbers are now more common than traditional European terms, the number is usually called a *billion* but is sometimes still named, according to the older system, a *thousand million*. In continental Europe it is usually called a *milliard*. A *centillion*, a number so large it cannot easily be written out in figures, is 10^{303} in the American system; in the traditional British system it is 10^{600}.

Number	American and modern British name	Traditional British and European name
1,000,000 (10^6)	million	million
1,000,000,000 (10^9)	billion	in Britain: thousand million; in continental Europe: milliard
1,000,000,000,000 (10^{12})	trillion	billion
1,000,000,000,000,000 (10^{15})	quadrillion	thousand billion
1,000,000,000,000,000,000 (10^{18})	quintillion	trillion
1,000,000,000,000,000,000,000 (10^{21})	sextillion	thousand trillion
1,000,000,000,000,000,000,000,000 (10^{24})	septillion	quadrillion
1,000,000,000,000,000,000,000,000,000 (10^{27})	octillion	thousand quadrillion
1,000,000,000,000,000,000,000,000,000,000 (10^{30})	nonillion	quintillion
1,000,000,000,000,000,000,000,000,000,000,000 (10^{33})	decillion	thousand quintillion

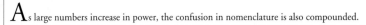 As large numbers increase in power, the confusion in nomenclature is also compounded.

A few strange number facts

Numbers have intriguing quirks, many of which can be made into puzzles and games. Hundreds of books of number problems and brain-teasers have been published. The following come from Numbers: Facts, Figures and Fiction, *by Richard Phillips, and* The Moscow Puzzles, *by Boris A. Kordemsky.*

• 2 is the only even prime number. There is no largest prime number.

• The number 9 has some interesting peculiarities: a number is divisible by 9 if the sum of its digits is divisible by 9. For example: 354 x 9 = 3,186 and 3 + 1 + 8 + 6 = 18 (divisible by 9).

• 504: if you subtract 7 from 504 the result is divisible by 7; if 8, divisible by 8; and if 9, divisible by 9.

• 9,801: $(98 + 01)^2 = 9,801$. This is true for two other four-digit numbers: $3,025 = (30 + 25)^2$ and $2,025 = (20 + 25)^2$.

• 1.496^{11}: 149,600,000,000 meters is the distance astronomers call *one astronomical unit* (au). It is the average distance between the earth and the sun.

• A light-year, another unit of astronomical measure, is the distance light travels in one year in a vacuum (as in space): 5,878,000,000,000 miles, or 9,460,500,000,000,000 meters. There are about 63,240 astronomical units in a light-year.

Against Pythagoras, against Zeno

The Greek Pythagoras (c. 580–c. 500 BC) and his followers saw the universe as not only harmonious, but harmonic. For them philosophy, religion, music, and mathematics were all aspects of a single, unified idea. This was one of the first great formulations of a philosophy of number in history. Appealing though it was, this system of thought attracted much opposition.

Zeno's paradoxes

The philosopher Zeno of Elea (mid-5th century BC) disputed the coherent and orderly Pythagorean view of the universe. An influential mathematician and logician, he is perhaps best known today for the paradoxes he proposed as tools with which to explore the nature of motion and to question the continuity of number and of space. Here, two modern scholars summarize them.

Four of them seem to have caused the most trouble: (1) the *Dichotomy,* (2) the *Achilles,* (3) the *Arrow,* and (4) the *Stade.* The first argues that before a moving object can travel a given distance, it must first travel half this distance; but before it can cover this, it must travel the first quarter of the distance; and before this, the first eighth, and so on through an infinite number of subdivisions. The runner wishing to get started must make an infinite number of contacts in a finite time; but it is impossible to exhaust an infinite collection, hence the beginning of motion is impossible. The second of the paradoxes is similar to the first except that the infinite subdivision is progressive rather than regressive. Here Achilles is racing against a tortoise that has been given a headstart, and it is argued that Achilles, no matter how swiftly he may run, can never overtake the tortoise, no matter how slow it may be. By the time that Achilles will have reached the initial position of the tortoise, the latter will have advanced some short distance; and by the time that Achilles will have covered this distance, the tortoise will have advanced somewhat farther; and so the process continues indefinitely, with the result that the swift Achilles can never overtake the slow tortoise.

The *Dichotomy* and the *Achilles* argue that motion is impossible under the assumption of the infinite subdivisibility of space and time; the *Arrow...*, on the other hand, argue[s] that motion is equally impossible if one makes the opposite assumption—that the subdivisibility of space and time terminates in indivisibles. In the *Arrow* Zeno argues that an object in flight always occupies a space equal to itself; but that which always occupies a space equal to itself is not in motion. Hence, the flying arrow is at rest at all times, so that its motion is an illusion.

Carl B. Boyer and
Uta C. Merzbach,
A History of Mathematics,
2d ed., 1991

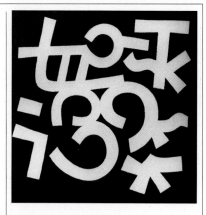

Numbers, according to the Pythagoreans, are not merely practical, but beautiful. Contemporary artists agree.

Aristotle responds to Pythagoras

The philosopher Aristotle (384–322 BC) acknowledged the debt that mathematics owed to the Pythagoreans. However, while he admitted the legitimacy of their study of numbers when applied to objects, he scorned their rigidly numerical conception of the world.

The so-called Pythagoreans, who were engaged in the study of mathematical objects, were the first to advance this study, and having been brought up in it, they regarded the principles of mathematical objects as the principles of all things. Since of mathematical objects numbers are by nature first, and (a) they seemed to observe in numbers, rather than in fire or earth or water, many likenesses to things, both existing and in generation (and so they regarded such and such an attribute of numbers as justice, such other as soul or intellect, another as opportunity, and similarly with almost all of the others), and (b) they also observed numerical *attributes*

and ratios in the objects of harmonics, since, then, all other things appeared in their nature to be likenesses of numbers, and numbers to be first in the whole of nature, they came to the belief that the elements of numbers are the elements of all things and that the whole heaven is a harmony and a number. And whatever facts in numbers and harmonies could be shown to be consistent with the *attributes,* the parts, and the whole arrangement of the heaven, these they collected and fitted into a system; and if there was a gap somewhere, they readily made additions in order to make their whole system connected. I mean, for example, that since ten is considered to be complete and to include every nature in numbers, they said that the bodies which travel in the heavens are also ten; and since the visible bodies are nine, they added the so-called "Counter-Earth" as the tenth body....

Indeed, these thinkers appear to consider numbers as principles of things, and in two senses: as matter and also as affections or possessions of things. The elements of a number are the *Even* and the *Odd,* the *Odd* being finite and the *Even* being infinite; the *One* is composed of both of these (for it is both even and odd); a number comes from the *One;* and, as we said, the whole heaven is numbers....

[But] from many numbers one number is formed, but how can one Form be formed from many Forms? And if it is not from them but from the Units in them that a Number is formed, such as 10,000 for example, how are the Units related to each other in the Number formed? Many absurdities will follow whether the Units (a) are all alike in kind, or (b) are not alike in kind, either in the sense that, prior to the formation of this Number, the Units of each Number are alike in kind but not alike in kind with those of any other Number, or in the sense that no one Unit is alike in kind with any other Unit. For, having no attributes, with respect to what will the Units differ? These alternatives are neither reasonable nor in agreement with our thinking.

Aristotle, *Metaphysics,*
Book A, chapter 5, 4th century BC,
trans. Hippocrates G. Apostle, 1966

A modern philosopher responds to Zeno

Bertrand Russell (1872–1970), the eminent English mathematician and philosopher, was deeply interested in logic and revisited Zeno's classic mathematical paradoxes.

The possibility that whole and part may have the same number of terms is, it must be confessed, shocking to common-sense. Zeno's Achilles ingeniously shows that the opposite view also has shocking consequences; for if whole and part cannot be correlated term for term, it does strictly follow that, if two material points travel along the same path, the one following the other, the one which is behind can never catch up: if it did, we should have, correlating simultaneous positions, a unique and reciprocal correspondence of all the terms of a whole with all the terms of a part. Common-sense, therefore, is here in a very sorry plight; it must choose between the paradox of Zeno and the paradox of [Georg] Cantor. I do not propose to help it, since I consider that, in the face of proofs, it ought to commit suicide in despair. But I will give the paradox of Cantor a form resembling that of Zeno. Tristram Shandy, [the hero of the eponymous novel,] as we know, took two years writing the history of the first two days of his life, and lamented that, at this rate, material would accumulate faster than he could deal with it, so that he could never come to an end. Now I maintain that, if he had lived for ever, and not wearied of his task, then, even if his life had continued as eventfully as it began, no part of his biography would have remained unwritten. This paradox, which, as I shall show, is strictly correlative to the Achilles, may be called for convenience the Tristram Shandy.

In cases of this kind, no care is superfluous in rendering our arguments formal. I shall therefore set forth both the Achilles and the Tristram Shandy in strict logical shape.

I. (1) For every position of the tortoise there is one and only one of Achilles; for every position of Achilles there is one and only one of the tortoise.

(2) Hence the series of positions occupied by Achilles has the same number of terms as the series of positions occupied by the tortoise.

(3) A part has fewer terms than a

whole in which it is contained and with which it is not coextensive.

(4) Hence the series of positions occupied by the tortoise is not a proper part of the series of positions occupied by Achilles.

II. (1) Tristram Shandy writes in a year the events of a day.

(2) The series of days and years has no last term.

(3) The events of the nth day are written in the nth year.

(4) Any assigned day is the nth, for a suitable value of n.

(5) Hence any assigned day will be written about.

(6) Hence no part of the biography will remain unwritten.

(7) Since there is a one-one correlation between the times of happening and the times of writing, and the former are parts of the latter, the whole and the part have the same number of terms....

These two paradoxes are correlative. Both, by reference to segments, may be stated in terms of limits. The Achilles proves that two variables in a continuous series, which approach equality from the same side, cannot ever have a common limit; the Tristram Shandy proves that two variables which start from a common term, and proceed in the same direction, but diverge more and more, may yet determine the same limiting class (which, however, is not necessarily a segment, because segments were defined as having terms beyond them). The Achilles assumes that whole and part cannot be similar, and deduces a paradox; the other, starting from a platitude, deduces that whole and part may be similar. For common-sense, it must be confessed, this is a most unfortunate state of things.

There is no doubt which is the correct course. The Achilles must be rejected, being directly contradicted by Arithmetic. The Tristram Shandy must be accepted, since it does not involve the axiom that the whole cannot be similar to the part. This axiom, as we have seen, is essential to the proof of the Achilles; and it is an axiom doubtless very agreeable to common-sense. But there is no evidence for the axiom except supposed self-evidence, and its admission leads to perfectly precise contradictions. The axiom is not only useless, but positively destructive, in mathematics, and against its rejection there is nothing to be set except prejudice. It is one of the chief merits of proofs that they instil a certain scepticism as to the result proved.

Bertrand Russell,
The Principles of Mathematics, 1903

P ythagoras, in a Renaissance woodcut, experiments with bells and water glasses to determine the numerical values of musical notes.

Numbers and religion

Numbering is the earliest form of measurement. In ancient times to take a census—which aims to measure humanity, to number the multitude—was sometimes considered a crime, or an act permitted only to God. "Who can count the dust of Jacob?" the Bible asks.

Biblical numbers

The fourth book of the Pentateuch is titled Numbers. There and in other places in the Old Testament the children of Israel are named and counted. Numbers, measurement, and counting often carry symbolic meanings in religious texts—of earthly or heavenly power, of divine or material bounty, or of trial and judgment.

And the Lord spake unto Moses....Take ye the sum of all the congregation of the children of Israel, after their families, by the house of their fathers, with the number of their names...all they that were numbered were six hundred thousand and three thousand and five hundred and fifty.

Numbers 1:1, 2:46,
King James Version

Yet the number of the children of Israel shall be as the sand of the sea, which cannot be measured nor numbered.
Hosea 1:10, King James Version

This is the writing that was written [on the wall], MENE, MENE, TEKEL, UPHARSIN. This is the interpretation of the thing: MENE; God hath numbered thy kingdom, and finished it. TEKEL; Thou art weighed in the balances, and art found wanting. PERES; Thy kingdom is divided.
Daniel 5:25–28, King James Version

Number in the *Dao De Jing*

An ancient Chinese religious text links number to spiritual harmony through the Dao, *or Way.*

Dao generates the One.
The One generates the Two.
The Two generates the Three.
The Three generates all things.

All things have darkness at their back
and strive towards the light,
and the flowing power gives them
    harmony.

> Lao Zi [Lao Tzu],
> *Dao De Jing [Tao Te Ching]*,
> 6th century BC,
> trans. H. G. Ostwald, 1985

A Hindu vision of infinity

In the Bhagavad-Gita, *or Song of God, a sacred Hindu epic written between the 5th and 2d centuries BC, the hero Arjuna has a vision of the infinite form of God.*

Ah, my God, I see all gods within your
    body;

Each in his degree, the multitude of
    creatures;
See Lord Brahma throned upon the
    lotus;
See all the sages, and the holy serpents.

Universal form, I see you without limit,
Infinite of arms, eyes, mouths and
    bellies—
See, and find no end, midst, or
    beginning....

You are all we know, supreme, beyond
    man's measure....

> *The Song of God: Bhagavad-Gita,*
> trans. Swami Prabhavananda and
> Christopher Isherwood, 1972

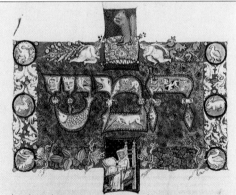

F ive as a holy number: a symbolic representation of the Hebrew word introducing the five books of Moses (the Pentateuch, or Torah), from a medieval manuscript.

Numbers, philosophy, and poetry

From earliest antiquity, numbers have been seen as representing deep and abstract ideas. Here, a classical and a modern philosopher think about numbers.

Plato's vision of mathematics

In The Republic, *a dialogue of Plato from the 4th century* BC, *the philosopher Socrates and his companion Glaucon explore the nature and value of mathematics as a branch of knowledge. What, Socrates asks,* "would be the study that would draw the soul away from the world of becoming to the world of being?" *They determine that neither gymnastics nor the arts can do this, but only a science that is universal and studies the truth.*

[Take] this common thing that all arts and forms of thought and all sciences employ, and which is among the first things that everybody must learn.

What? [Glaucon] said.

This trifling matter, I said, of distinguishing one and two and three. I mean, in sum, number and calculation. Is it not true of them that every art and science must necessarily partake of them?

Indeed it is, he said....

But, further, reckoning and the science

Plato as depicted in the Renaissance

of arithmetic are wholly concerned with number.

They are, indeed.

And the qualities of number appear to lead to the apprehension of truth.

Beyond anything, he said.

Then, as it seems, these would be among the studies that we are seeking. For a soldier must learn them in order to marshal his troops, and a philosopher because he must rise out of the region of generation and lay hold on essence or he can never become a true reckoner....

And, further, I said, it occurs to me, now that the study of [numbers] has been mentioned, that there is something fine in it, and that it is useful for our purpose in many ways, provided it is pursued for the sake of knowledge and not for huckstering.... [For] it strongly directs the soul upward and compels it to discourse about pure numbers, never acquiescing if anyone proffers to it in the discussion numbers attached to visible and tangible bodies. For you are doubtless aware that experts in this study, if anyone attempts to cut up the "one" in argument, laugh at him and refuse to allow it, but if *you* mince it up, *they* multiply, always on guard lest the one should appear to be not one but a multiplicity of parts.

Most true, he replied.

Suppose now, Glaucon, someone were to ask them, My good friends, what numbers are these you are talking about, in which the one is such as you postulate, each unity equal to every other without the slightest difference and admitting no division into parts? What do you think would be their answer?

This, I think—that they are speaking of units which can only be conceived by thought, and which it is not possible to deal with in any other way.

You see, then, my friend, said I, that this branch of study really seems to be indispensable for us, since it plainly compels the soul to employ pure thought with a view to truth itself.

Plato, *The Republic,* Book 7, trans. Paul Shorey, in *Plato: The Collected Dialogues,* 1989

The epic saga of numbers

A literary critic, concerned with the sublime nature of poetry, finds himself drawn to the study of number and counting, and considers the meaning of both.

There is no more impressive form of literature than the narrative epic poem. That combination of depth and breadth of conception which some have called sublimity has here found a natural and adequate expression. The theory of number is the epic poem of mathematics. The mutual reflection of the two arts will supply a sort of explanation in the intellectual dimension of the epic quality in both. It will also show how it is that the number has made some of the discoveries in physical science possible; it is to be remarked that these physical discoveries themselves have an epic quality probably due to the part numbers have played in their technical development.

But the question, what is a number? is an invitation to analyze counting and find out what sort of thing makes counting possible. In poetry I suppose the corresponding question would be, what makes recounting possible? The answer, if it were to be complete, would take us into the most abstract and subtle mathematical thought. But the key to the problem is the simplest sort of insight. The same peculiar combination of simplicity and subtlety is involved in the theory of narrative, but as everyone knows, insight here belongs to the most common of common-sense conceptions.

Very briefly, a number is an element

in a field of variation. Its specific numerical property consists in its relation with other elements in the same field, which relations are expressed in rules of order. One can see immediately how a significant event or incident in a story conforms to this definition in the general form I have given it. In geometry, the emphasis is on the constancy of elements that undergo transformation; the emphasis in arithmetic is on the order and connection of the elements in the transformation itself. Arithmetic adds the "how" to geometry's "what." Numbers reveal another aspect of the mathematical object.

The structures that make counting possible are chains and networks of relations. The understanding of a selected few of these relations will be the suggestive key to the whole realm. There are two ways of approaching these relations. One has the advantage of being itself a kind of narrative that gathers up the main turning points in the history of arithmetic. This is the theory of operations. The other is postulate theory in which collections of elements are assumed and one relation after another is introduced until a mere aggregate becomes as if by magic a number system. I shall begin with the theory of operations. It is merely a careful account of counting with special attention paid to the difficulties met and solved or circumvented.

A point of beginning is assumed, usually the number one. This is the fulcrum demanded by Archimedes when he said he would move the world by means of levers. An operator, so called, is also taken—one of Archimedes' levers. This operator is called +1. Then there follows a rapid process of intellectual knitting starting with $1 + 1 = 2$, $2 + 1 = 3$, $3 + 1 = 4,\ldots$ and ending (?) with $n + 1$, and you have the series of positive

numbers to infinity. Of course infinity is not a number in this series but "to infinity" describes a property of the series, namely, that it has only the arbitrary end you wish to assign it if you want a finite series; or it has no end, that is, no last term.

Then you take another operator, -1, and starting with any one of the positive numbers, you can travel backwards along this series until you arrive at 1. Here you encounter a difficulty, the solution of which once bothered the mathematical consciences of many men, as much as warped space does now. What can $1 - 1$ mean? The answer is another symbol, unknown before, and, in a sense, unknown still. It is one of those vehicles which one can ride without looking under the hood. Of course we know it is 0....

Between any two numbers in [the series of real numbers] it is possible to find a third, no matter how many you have introduced in that way before. It has no beginning and no end, and in a peculiar way Zeno's paradox about passing over an infinite number of points in finite time is given further exemplification even without points and time. You can never pass from one number to another by any thoroughly step-by-step procedure. It is now a riddle not only how bodies move, but also a much worse one, how anybody counts. Incidentally, in showing how we count, we have shown how counting is impossible.

It would seem that the people with mathematical consciences were wise in sticking to the positive integers and that these riders of newfangled vehicles have gone too far. On another occasion very much like this Bishop Berkeley accused mathematicians of dealing in symbols more vicious and unintelligible than those of theology, and others went on to show that if mathematicians could prove eternal truths by such obviously questionable methods, how much more right had mystics and preachers to use questionable arguments for which they made no such ambitious claims.

Obviously here is a tangle, but it is easy to unravel. Counting is not covering ground, any more than measuring a distance or telling a story is covering ground. It is a little more subtle. Counting has always to do with at least two sets of numbers, what have been called the numbering numbers and the numbered numbers, and the process is called one-oneing or correlation, that is, finding in one of two series a corresponding number for each number in the other. It is very easy to find a series of numbered numbers for the series of fractional or rational numbers. Euclidean space is such a series and in addition contains some extra numbers called irrationals....The paradox of counting by means of an infinite series is only the occasion for new discoveries about the network of numbers. Operations merely outran analysis for a time, but like the tortoise, analysis catches up, and when it does, one sees that the conditions of the whole race are revised. Counting is a different thing because mathematicians played with unintelligible operations for a time.

But we must shift our ground of explanation of numbers to see just what this amounts to. The shift is like a certain one in narrative literature that happened long ago.... [At one time a] story was more than a story. It bore a burden, sometimes an insight, and sometimes even a moral. Epic literature has often become sacred scripture and, even more often, the repository of a people's history and civilization. When it has done this, it has not been merely a record of events and gossip about them. It has been interpretive, as we used to say. It has had a certain generality or universality....I suggest that that something is the relations holding between the events or incidents. In other words, the plot of a story is an intellectual as well as an aesthetic pattern and it is this that gives the incidental elements that float in it significance. These patterns are very abstract, very general, and capable of infinite variation, so that they may be revised and reapplied without violating their essential forms....

I prefer to compare [story] with number, that age-old story recounted at one time on fingers and toes, at another in the knots of a rug, at another in the constellations, the letters of the alphabet, the beads of wampum, and the modern cash register of a ten-cent store. The affinity of stories and numbers is always latent, bursting forth in magic formulae, sacred numbers, astrology, alchemy, the Cabala, and the wisdom of the Rosicrucians, and periodically hypostasized into the supreme dogma of a universal religion, as in the Trinity or the infinitude of deities in Eastern religions. Numbers are not just counters; they are elements in a system. It is this aspect of numbers with which the story of postulate theory concerns itself. It is in the findings of such studies that the necessary conditions of counting are stated.

These results have come from a peculiar sort of study. The modern mathematician has been sitting down like an Epicurean god, far from space and time, calling for chaos to play with. He gives a few brief orders and watches universes grow.

Scott Buchanan,
Poetry and Mathematics, 1962

The science of measurement

Measurement, which quantifies phenomena, was once the most essential tool of the sciences, largely giving them their operational power. Its importance in both the history of research and general history is undisputed. Nowadays, however, science concerns itself more and more with the nonquantitative aspects of phenomena and the "all-powerful measure" no longer reigns supreme.

The decimal revolution

In France the uniform decimal metric system was the child of the French Revolution, which sought to reform many antiquated and inconsistent public structures. Weights and measures were all to be calculated in tens, and the calendar and clock were reorganized as well. In order to establish the universality of the metric system, multiples were given classical Greek and Latin prefixes still in use today: hecto-, kilo-, deci-, centi-, milli-.

The decimal system was not popularized easily or completely; to promote it many public notices were published. This letter from a French revolutionary regional government office is addressed to a professor of mathematics.

Citizen,
The Revolution not only improves our morals and paves the way for our happiness and that of future generations, it even unlooses the shackles that hold back scientific progress. Our arithmetic, one of the greatest inventions of the human mind, long remained subject to the tyrannical yoke of our old gothic and barbarian laws. In vain the great inventors of this science had based it on the simple and fruitful principle that once…[numbers are determined and fixed, they will follow a uniform system]. This great principle of numeration could have been applied to all magnitudes and

A Greek bas-relief in which a man's armspan indicates a unit of measure.

quantities, but was applied only to abstract magnitudes, while [until the Revolution] our absurd institutions used fractions that did not follow the numerical rules at all: *sous* were coins worth one-twentieth of a pound; *deniers,* smaller coins, were worth one-twelfth of a *sou,* and so on. It was an absurd system of currency.

The system of weights and measures proved no less absurd. The pound-weight was subdivided into halves by *marcs;* the *marcs* into eighths by ounces; and so on.

Time, that abstract entity that seemed subject only to the empire of mathematics, had nevertheless become the slave of tyrannical custom, which had enslaved us as well. The year was divided into 365 days and some hours, the days into 24 hours, the hours into minutes, minutes into seconds, and seconds into sixty units, all following the sexagesimal system.

We will no longer suffer these contradictions, citizen Professor: we join you in congratulating ourselves that the genius of the Revolution has overturned all the customs born in the shadows and set in their place the simple and methodical *decimal* system. The time approaches when this system will be in common usage. However simple it may be, it must neverthe-less be taught; we must set aside our old routine and become acquainted with and adapt to the new method. It is up to teachers like you, who combine theory and practice, to start the process, to instruct their fellow citizens. We thus commend the enthusiasm and republicanism that have led you to start a class in republican arithmetic.

You express the desire that all citizens be able to take advantage of it, especially those who work in government. You realize, citizen, that those who work in government keep hours from eight in the morning (old style) until four in the afternoon. We must have our dinners! Your course cannot be useful to anyone employed in government unless you do not begin your lessons until between five and six o'clock. We are confident that, barring some major obstacle, you will choose that hour.

You may count on the regard and

ARIDMETRICA

A Renaissance allegory of Arithmetic.

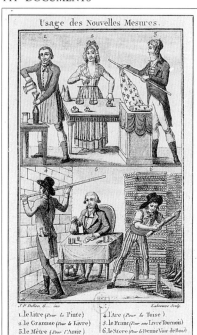

Usage des Nouvelles Mesures.

1. le Litre (*Pour la Pinte*)   4. l'Are (*Pour la Toise*)
2. le Gramme (*Pour la Livre*)   5. le Franc (*Pour une Livre Tournois*)
3. le Mètre (*Pour l'Aune*)   6. le Stere (*Pour la Demme Voie de Bois*)

J.P Delion G..... inv   Laboureur Sculp.

A Paris chez Delion Rue Montmartre N°no près le Boulevard.

A French illustration of about 1796 demonstrates the new system of measurement.

---

esteem of all good citizens, as long as your efforts have only the prosperity of the Republic as their goal.

Greetings and brotherhood.

Letter from the Revolutionary Directorate of the Seine-Inférieure to Caius Gracchus Prudhomme, 17 March 1794

The power of measurement

The modern French poet and critic Paul Valéry (1871–1945) was greatly interested in the sciences. In the following passage he considers the commercial power of measurement, and the way in which it links pure mathematics to applied, or practical, uses.

No place is more favorable, no environment more stimulating to meditation on the grand design…than this center of trade, where *measure* reigns supreme. Everything in a [sea]port is manifestly, openly, brutally metrical. The whole of its visible activity is concentrated on counting, weighing, classifying, and stowing away; number and order plainly dictate every action, and nothing passes through it which cannot be gauged in terms of tons, pounds, bushels, and sundry other measures.

Is not the Method, after all, the Charter of a realm of Number whose whole ambition is now apparent to us, even if we cannot yet grasp its full power? The measurable has conquered almost the entire field of the sciences and has discredited every branch in which it is not valid. The applied sciences are almost completely dominated by measurement. Life itself, which is already half enslaved, circumscribed, streamlined, or reduced to a state of subjection, has great difficulty in defending itself against the tyranny of timetables, statistics, quantitative measurements, and precision instruments, a whole development that goes on reducing life's diversity, diminishing its uncertainty, improving the functioning of the whole, making its course surer, longer, and more mechanical.

Paul Valéry,
Masters and Friends,
trans. by Martin Turnell, 1968

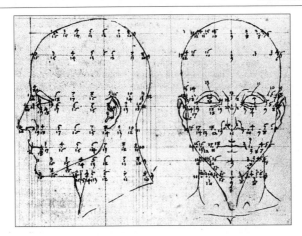

Detailed measurements of the human head by the great Renaissance painter and mathematician Piero della Francesca (c. 1420–92).

A professor watches his student work with an abacus in Japan at the turn of the century. The modern science of measurement relies on the ancient science of calculation.

The abacus and the calculator

Since paleolithic times, when protomathematicians tallied numbers in notches on bones, people have invented tools to help them calculate.

The evolution of the calculator

The electronic calculator comes from a long line of earlier, hand-powered mechanical instruments. Calculators and computers of ever greater power and sophistication continue to be developed.

The first hand-held electronic calculator was created by three American engineers at the Texas Instruments company; it was a simple machine that could add, subtract, multiply, and divide, just like the more sophisticated mechanical calculators at that time. It did, though, manage to do one thing its mechanical peers could not: replace the abacus and the slide rule as the preferred means of solving problems.

Today's devices, of course, are a quantum leap ahead of that early model. They have graphical displays and can perform a wide range of complex scientific and mathematical functions....But, while the hand-held calculator has grown up, it is no longer necessarily the tool of choice....Often, computers are now equipped with software that can do the same things as their hand-held cousins. The World Wide Web [of the Internet], too, offers a number of virtual calculators able to do everything from computing sine waves to figuring out mortgage payments. And the abacus, which is still common in nonindustrialized countries, is even making a comeback as a teaching tool in some schools....

Which system will people turn to most often to balance their checkbooks or

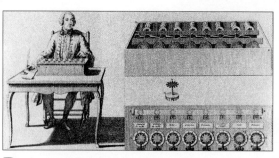

Blaise Pascal calculating with his 17th-century computer.

calculate the trajectory of a missile? Only time will tell. But then, who could have foreseen how the technology would develop in the first place? Here is a guide to the development of the calculator.

Abacus

Invented between 2300 BC and 500 BC, the origins of the archetypal calculator are uncertain. There are several kinds. A Chinese suanpan is identified by its five beads below the bar and two above on each rod. The Russian form has no center bar. Use of the abacus has persisted into this century. The Russians, for example, used it [see page 56] to do some of the calculations for the launch of Sputnik [the first satellite] in 1957.

Slide rule, 1621

Shortly after logarithms were invented, the English mathematician Edmund Gunter (1581–1626) plotted a number of them on 2-foot-long wooden rules, along with other specialized functions that abacuses could not handle. The next year William Oughtred (1574–1660), also English, joined rules together to create the first slide rule. Slide rules remained very popular well into the 1960s and early 1970s....

First mechanical calculator, 1642

Invented by Blaise Pascal (1623–62), it used gears to add a column of up to eight figures, but was expensive to build and maintain.

Difference engine, 1833

Invented by Charles Babbage (1792–1871), only one-seventh the functions he had intended ever worked. But it showed how calculations more complex than addition, subtraction, multiplication, and division could be handled mechanically.

Mark I, 1944

Using the lessons of Babbage's machine, scientists at the IBM corporation created the first electromechanical computer, the Automatic Sequence Controlled Calculator, or Mark I. Within two years, though, it was obsolete.

Eniac, 1946

A more powerful and faster computer, the Electronic Numerical Integrator and Computer, or Eniac, was designed by re-searchers at the University of Pennsylvania. It relied on thousands of vacuum tubes, resistors, capacitors, and switches, and it was housed in a 30-foot-by-50-foot room.

Hand-held calculator, 1967

The development of the integrated circuit paved the way for this machine, though at first it had little more power than its mechanical predecessors.

Today's calculators, 1997

Hand-held models allow the user to graph the results of calculations. On the World Wide Web, virtual calculators simulate the format of the actual versions.

Dylan Loeb McClain,
The New York Times,
1 September 1997 (revised)

The German mathematician and philosopher Gottfried Wilhelm Leibniz, pioneer in the fields of logic and calculus, also designed a multiplying machine in 1683.

Music and mathematics

Music and mathematics both express abstract ideas in a special language of notation that is internationally understood. But the relationship goes deeper. A plucked string produces a sound. The pitch of the sound depends on the length of the string. The Pythagoreans were perhaps the first to explain this through numbers. They discovered that the intervals of the octave, the fifth, and the fourth are in the ratios of $\frac{1}{2}$, $\frac{2}{3}$, and $\frac{3}{4}$— three of the simplest fractions. By establishing a link between two disciplines, music and mathematics, the Pythagoreans revealed a new way of knowing nature.

Number as a principle of reality

The Pythagorean Philolaus of Croton (470–390 BC) expresses the sect's religious view of numbers and music.

It is in the nature of number that the possibility of recognition resides; it gives direction and teaching to every man with respect to what is unknown and baffling. Nothing about existing things—neither they themselves nor their relations to one another—would be clear without number and the essence of number. It is number which takes things that we apprehend by sense-perception and fits them harmoniously into the soul, thereby making them recognizable and capable of being compared with one another, as the power of the gnomon makes possible. Thereby it gives body to things and distinguishes the different relations between things, whether unlimited or limiting. You can see the nature and power of number illustrated not only in spiritual and divine matters but also [implicitly] in human affairs and in language quite generally, including productive activities in all the crafts as well as in music.

Falsehood does not inhere in the nature of number and harmony; for there is no kinship between it and them. Falsehood and envy partake of the nature of the unlimited, the unreasonable, and the irrational. Falsehood cannot be breathed into number, being hostile and inimical to its very nature; whereas truth is congenial to number and shares close family ties with it.

Philolaus, *On the World,*
5th century BC,
from Stobaeus,
Anthology I, xxi, 7–8,
in *The Presocratics,*
trans. Philip Wheelwright, 1966

The laws of harmony

The French Baroque composer Jean-Philippe Rameau (1683–1764) was accomplished in both science and music. In his theoretical writing he was passionately interested in linking the two, evoking Pythagoras and the ancient tradition of arithmetic that held that numbers contained the key to the soul of the world. He wrote one of the first scientific analyses of musical harmony.

Music is a science which should have definite rules; these rules should be drawn from an evident principle; and this principle cannot really be known to us without the aid of mathematics.

Notwithstanding all the experience I may have acquired in music from being associated with it for so long, I must confess that only with the aid of mathematics did my ideas become clear and did light replace a certain obscurity of which I was unaware before....I could not help thinking that it would be desirable (as someone said to me one day while I was applauding the perfection of our modern music) for the knowledge of musicians of this century to equal the beauties of their compositions. It is not enough to feel the effects of a science or an art. One must also conceptualize these effects in order to render them intelligible. That is the end to which I have principally

Detail of a sonata, showing the time signatures.

applied myself in the body of this work, which I divided into four books.

The First Book contains a summary of the relationship between sounds, consonances, dissonances, and chords in general. The source of harmony is discovered to be a single sound and its most essential properties are explained. We shall see, for example, how the first division of this single sound generates another sound, which is its octave and seems to be identical to the first sound, and how the latter then uses this octave to form all the chords. We shall see that all these chords contain only the source, its third, its fifth, and its seventh, and that all the diversity inherent in these chords derives from the power of the octave. We shall discover several other properties, perhaps less interesting for practice but nonetheless necessary for achieving proficiency. Everything is demonstrated in the simplest manner....

In order to understand the relationship between sounds, investigators took a string, stretched so that it could produce a sound, and divided it with movable bridges into several parts. They discovered that all the sounds or intervals that harmonize were contained in the first five divisions of the string, the lengths resulting from these divisions being compared with the original length.

Some have sought an explanation of this relationship in that relationship existing between the numbers indicating the [number of] divisions. Others, having taken the lengths of string resulting from these divisions, have sought an explanation in the relationship between the numbers measuring these different lengths. Still others, having further observed that communication of sound to the ear cannot occur without the participation

of the atmosphere, have sought an explanation in the relationship between the numbers indicating the vibrations of these various lengths. We shall not go into the several other ways in which this relationship may be known, such as with strings of different thicknesses, with weights which produce different tensions in the strings, with wind instruments, etc. It was found, in short, that all the consonances were contained in the first six numbers, except for the methods using thicknesses and weights, where the squares of these fundamental numbers had to be used. This has led some to attribute all the power of harmony to that of numbers; it is then only a matter of applying properly the operation on which one chooses to base one's system.

We must remark here that the numbers indicating the divisions of the string or its vibrations follow their natural progression, and that is based on the rules of arithmetic.

Jean-Philippe Rameau,
Treatise on Harmony, 1722,
trans. Philip Gossett, 1971

Beauty and the sublime

The contemporary critic Edward Rothstein has written in philosophical terms of the deep relationship between music and mathematics, particularly the ways in which both arts (as he calls them) pursue a certain ideal of beauty.

[Henri] Poincaré wrote, "The Scientist does not study nature because it is useful to do so. He studies it because he takes pleasure in it; and he takes pleasure in it because it is beautiful."

But this judgement of beauty in mathematics or the natural world seems far more difficult to understand than the processes of reasoning that bring us to such a judgment. Judging a formula, an insight, a method to be beautifully constructed is asserting that the mathematical argument has a quality or purpose quite apart from the reasoning processes that gave it birth. It is as if we were judging it the way the Deity judged His own creations after each day of labor: it is good. We don't need to assert divine intervention in order to experience this wonder. What is important is that at the very moment such beauty becomes apparent to us, the world appears *as if* it were constructed according to some purpose. Or better: the world appears as if it were well suited to human perception and understanding, as if nature were constructed specifically for our contemplation. [The philosopher Immanuel] Kant referred to this as regarding nature "after the analogy of art."

It might seem peculiar to be bringing in such indirect and imprecise sentiments in talking about a subject [mathematics] which insists upon clarity and precision. But these feelings are related to the reasons theories are constructed in the first place. A theory must create order, define conditions, show the relationship between parts and whole. It explains things, but it also discovers things, revealing unexpected connections. "You must have felt this too," the great physicist Werner Heisenberg said to Albert Einstein, "the almost frightening simplicity and wholeness of the relationships which nature suddenly spreads out before us and for which none of us was in the least prepared."

The mathematician J. W. N. Sullivan wrote, "Since the primary object of the scientific theory is to express the harmonies which are found to exist in nature, we see at once that these theories must have an aesthetic value. The measure of the success of a scientific theory is, in fact, a measure of its aesthetic value, since it is a measure of the extent to which it has introduced harmony in what was before chaos." Further: "The measure in which science falls short of art is the measure in which it is incomplete as science."

…The search for the sublime links music and mathematics. Both arts seek something which combined with the beautiful provokes both contemplation and restlessness, awe and comprehension, certainty and doubt. The sublime in mathematics and music sets the mind in motion, causes it to reflect upon itself. We become aware first, in humility, of the immensity of the tasks of understanding before us and the inabilities of human imagination to encompass them. The sublime inspires an almost infinite desire, a yearning for completion which is always beyond our reach. But we are then comforted by the achievements of reason in having brought us so close to comprehending a mystery fated to remain unsolved.

Edward Rothstein,
*Emblems of Mind:
The Inner Life of Music
and Mathematics,* 1995

Number and psychology

How do we first learn about numbers? At what age do we begin to understand what they mean? When is the idea of number born in the mind? Once we have grasped the concept of number, how do we develop ideas of logic and proof? Numbers lead us to highly disciplined ways of thinking.

How children become numerate

The innovative Swiss psychologist Jean Piaget (1896–1980) observed that the mind does not have an innate or intuitive understanding of the concept of number, but must learn it. Though the idea of number is learned by all children, no individual can recall later just when the knowledge arrived.

The psychological difficulty of the proposition of a primitive number intuition is that the series of numbers characterized by the operation $n + 1$ is discovered only bound up with the constitution of category and relation operations. On the preoperatory level (before the age of six or seven), although the child is unable to constitute the invariants needed for reasoning, for lack of reversible operations, he is well able to constitute the first numbers, which may be called figural because they correspond to simple and definite spatial arrangements (from one to five or six, without the zero), likewise he reasons by preconcepts corresponding to intuitive collections. But even in regard to groups of five or six objects, he is not certain of their conservation. When, for example, we ask a child of four or five to place on the table as many red chips as there are in a row of six blue chips spaced far apart, he begins by making a row of the same length, independent of the end to end correspondence. Then he forms a row with exact correspondence; but he is still basing himself on an exclusively perceptive criterion; he places each red chip in regard to the corresponding blue

The brilliant gaze of an infant expresses the astonishment of a mind grasping and absorbing new concepts of great complexity and abstraction with dazzling speed.

one, but if we move apart or together even a little the elements of one of the two rows, he no longer believes in the conservation of the equivalence and imagines that the longest row contains more elements. It is only when the child is six and a half or seven, that is, in connection with the formation of other conservation notions, that he will admit the invariance of everything irrespective of spatial position. Thus it is difficult to mention a whole number intuition before this last level. It is clear that an intuition which is not primitive is no longer an intuition!

How then are equivalence between two collections and conservation of this equivalence constituted? Logical operations here necessarily intervene and seem to prove [Bertrand] Russell's proposition correct. It is indeed remarkable that construction of whole number series is made precisely at the intellectual level (six to seven years of age) where these two principal structures of the qualitative logic of categories and relations are constituted: first, the interlocking system by inclusion, the basis of classification (categories A and A' are included in B, B and B' in C, and so forth); second, linkage or seriation of transitive asymmetrical relations (A smaller than B, B smaller than C, and so forth). The first of these two structures intervenes precisely in the conservation of groups; indeed conservation of a whole supposes a set of hierarchic inclusions joining to this whole the parts of which it is formed. As for seriation, it intervenes in the numerical order of the elements and psychologically constitutes one of the conditions of setting up a correspondence....

A child's alphabet and number chart from the turn of the century.

When a child draws a funny figure in reference to a model, he makes the parts of his drawing correspond to those of the model: a head corresponds to a head, a left hand to a left hand, without these elements being interchangeable. Thus there is here a qualitative correspondence, each element characterized by definite qualities, though we cannot mention a general unity. On the contrary, when the same child makes six red chips correspond to six blue chips, any one of the second can correspond to any one of the first on the condition that there is end to end correspondence. Correspondence thus becomes "general" since there is disregard of qualities, and the elements thus stripped of their distinctive characteristics are transformed into interchangeable units.

When the logician tells us that the category of Napoleon's marshals is equivalent to that of the signs of the zodiac and of the apostles, the category of all these categories being the "category of equivalent categories" which constitutes the number 12, is it a question of a "qualitative" or of a "general" correspondence? It goes without saying that it is general. There are no common qualities between Marshal Ney, Saint Peter, and the zodiacal sign Cancer; the elements of each category correspond to those of the other categories as interchangeable units and after abstraction of their qualities.

Psychologically, the explanation of the cardinal number by category operations is based therefore on a vicious circle; people mention a category of equivalent categories as though their equivalence resulted from their nature as categories, whereas we began by brushing to one side the "qualitative" correspondence (which alone stems directly from the nature of logical categories) for the benefit of a "general" correspondence, without noticing that this already transforms by itself the qualified individual elements of the category into numerical units. We have therefore transformed category into number but by introducing number from without by means of the "general" correspondence.

Actually, the whole number is really a product of logical operations…, but it combines the operations among them in an original manner which is irreducible to pure logic, and thus we must turn to a third solution….

Let there be a group of elements A, B,

and C, and so forth. If the subject is interested in their qualities, he can first begin by classifying them in various manners, which means gathering them according to their resemblances (or differences) but independent of order (if A equals B, the one neither precedes nor follows the other), or else he can arrange them according to their order of size or position, and so forth, yet ignoring their resemblance. In the first case, group elements are thus gathered as equivalences and in the second as differences, but elementary logical operations do not allow for connecting two objects simultaneously as equivalents (category) and differences (order relation). To transform these logical operations into numerical operations consists, on the contrary, in disregarding qualities and, consequently, considering two general elements of the category as both equivalent to all (1 = 1) and yet distinct: distinct because their enumeration, however the order chosen, always supposes for lack of any other distinctive characteristic that one is designated before the other or after it. The whole number is therefore psychologically a synthesis of category and of transitive asymmetrical relation, that is, a synthesis of logical operations yet coordinated among them in a new manner because of elimination of distinctive qualities. That is why in the finite any whole number simultaneously implies a cardinal and an ordinal aspect.

Jean Piaget,
Psychology and Epistemology,
trans. by Arnold Rosin, 1971

Inductive reasoning

Among the different types of proofs, the inductive proof, which reasons from particular cases to general ones, determines the properties of the natural numbers. For

Henri Poincaré at his desk.

example, let S be a class. Suppose that zero belongs to S and that for each individual number that belongs to this class, the one that follows it also belongs. Hence, all the numbers belong to this class. Aristotle established the rules of inductive and deductive logic in the Posterior Analytics *and other texts. Henri Poincaré, in the modern era, pursued this concept of reasoning. An aspect of inductive logic is the principle of reasoning by recurrence, which works in the following way:*
1) prove that a property is true for n;
2) supposing that it is true for n, *prove that it is true for the number that follows,* n + 1. *This done, we may conclude that the property is true for all the integers.*

The essential characteristic of reasoning by recurrence is that it contains, condensed, so to speak, in a single formula, an infinite number of syllogisms. We shall see this more clearly if we enunciate the syllogisms one after another. They follow one another, if one may use the expression, in a cascade. The following are the hypothetical syllogisms:—The theorem is true of the number 1. Now, if it is true of 1, it is true of 2; therefore it is true of 2. Now, if it is true of 2, it is true of 3; hence it is true of 3, and so on. We see that the conclusion of each syllogism serves as the minor of its successor. Further, the majors of all our syllogisms may be reduced to a single form. If the theorem is true of $n - 1$, it is true of n.

We see, then, that in reasoning by recurrence we confine ourselves to the enunciation of the minor of the first syllogism, and the general formula which contains as particular cases all the majors. This unending series of syllogisms is thus reduced to a phrase of a few lines.

Henri Poincaré,
Science and Hypothesis, 1905,
trans. J. Larmor, 1952

The wit and wisdom of numbers

Dry, technical, intimidating: this is how, too often, we think of numbers. Yet some of the cleverest humorists in the world have relied on mathematics to challenge, amuse, provoke, and delight. For we acknowledge that numbers dwell in a separate universe, yet it is one we can all enter. From the sublime to the ridiculous, numbers offer entertainment.

Numerical worlds

What do a great Irish satirist of the 18th century, two Victorian Englishmen (one a shy Oxford mathematician and other a sober Shakespeare scholar), and a modern American children's author have common? They all created wonderful fantasy lands in which numbers rule—for better or worse! Here are four puzzling places…

At my alighting I was surrounded by a Crowd of People, but those who stood nearest seemed to be of better Quality. They beheld me with all the Marks and Circumstances of Wonder; neither indeed was I much in their Debt; having never till then seen a Race of Mortals so singular in their Shapes, Habits, and Countenances. Their Heads were all reclined to the Right, or the Left; one of their Eyes turned inward, and the other directly up to the Zenith. Their outward Garments were adorned with the figures of Suns, Moons, and Stars, interwoven with those of Fiddles, Flutes, Harps, Trumpets, Guittars, Harpsicords, and many more Instruments of Musick, unknown to us in *Europe*. I observed here and there many in the Habit of Servants, with a blown Bladder fastned like a Flail to the End of a short Stick, which they carried in their Hands. In each Bladder was a small Quantity of dried Pease, or little Pebbles, (as I was afterwards informed). With these Bladders they now and then flapped the Mouths and Ears of those who stood near them, of which Practice I could not then conceive the Meaning. It seems, the Minds of these People are so taken up with intense Speculations, that they neither can speak, or attend to the Discourses of others, without being rouzed by some external Taction upon the Organs of Speech and Hearing; for which Reason, those Persons who are able to afford it, always keep a *Flapper,* (the Original is *Climenole*) in their Family, as one of their Domesticks; nor even walk abroad or make Visits without him. And the Business of this Officer is, when two or more Persons are in Company, gently to strike with his Bladder the Mouth of him who is to speak, and the Right Ear of him or them to whom the Speaker addresseth himself. This *Flapper* is likewise employed diligently to attend his Master in his Walks, and upon Occasion to give him a soft Flap on his Eyes; because he is always so wrapped up in Cogitation, that he is in manifest Danger of falling down every Precipice, and bouncing his Head against every Post; and in the Streets, of jostling others, or being jostled himself into the Kennel....

We entered the Palace, and proceeded into the Chamber of Presence; where I saw the King seated on his Throne, attended on each Side by Persons of prime Quality. Before the Throne, was a large Table filled with Globes and Spheres, and Mathematical Instruments of all Kinds. His Majesty took not the least Notice of us, although our Entrance were not without sufficient Noise, by the Concourse of all Persons belonging to the Court. But, he was then deep in a Problem, and we attended at least an Hour, before he could solve it. There stood by him on each Side, a young Page, with Flaps in their Hands; and when they saw he was at Leisure, one of them gently struck his Mouth, and the other his Right Ear; at which he started like one awaked on the sudden, and looking towards me, and the Company I was in, recollected the Occasion of our coming, whereof he had been informed before. He spoke some Words; whereupon immediately a young Man with a Flap came up to my Side, and flapt me gently on the Right Ear; but I made Signs as well as I could, that I had no Occasion for such an Instrument; which as I afterwards found, gave his Majesty and the whole Court a very mean Opinion of my Understanding. The King, as far as I could conjecture, asked me several Questions, and I addressed my self to him in all the Languages I had. When it was found, that I could neither understand nor be understood, I was conducted by his Order to an Apartment in his Palace, (this Prince being distinguished above all his Predecessors for his Hospitality to Strangers,) where two Servants were appointed to attend me. My Dinner was brought, and four Persons of Quality, whom I remembered to have seen very near the King's Person, did me the Honour to dine with me. We had two Courses, of three Dishes each. In the first Course, there was a Shoulder of Mutton, cut into an Æquilateral Triangle; a Piece of Beef into a Rhomboides; and a Pudding into a Cycloid. The second Course was two Ducks, trussed up into the Form of fiddles; Sausages and Puddings resembling Flutes and Hautboys, and a Breast of Veal in the shape of a Harp. The Servants cut our Bread into Cones, Cylinders, Parallelograms, and several other Mathematical figures. While we were at Dinner, I made bold to ask the Names of several Things in their Language; and those noble Persons, by the Assistance of their Flappers, delighted to give me Answers, hoping to raise my Admiration of their great Abilities, if I could be brought to converse with them. I was soon able to call for Bread, and Drink, or whatever else I wanted....

Those to whom the King had

entrusted me, observing how ill I was clad, ordered a Taylor to come next Morning, and take my Measure for a Suit of Cloths. This Operator did his Office after a different Manner from those of his Trade in *Europe*. He first took my Altitude by a Quadrant, and then with Rule and Compasses, described the Dimensions and Out-Lines of my whole Body; all which he entred upon Paper, and in six Days brought my Cloths very ill made, and quite out of Shape, by happening to mistake a Figure in the Calculation. But my comfort was, that I observed such Accidents very frequent, and little regarded....

The Knowledge I had in Mathematicks gave me great Assistance in acquiring their Phraseology, which depended much upon that Science and Musick; and in the latter I was not unskilled. Their Ideas are perpetually conversant in Lines and Figures. If they would, for Example, praise the Beauty of a Woman, or any other Animal, they describe it by Rhombs, Circles, Parallelograms, Ellipses, and other Geometrical Terms; or else by Words of Art drawn from Musick, needless here to repeat. I observed in the King's Kitchen all Sorts of Mathematical and Musical Instruments, after the Figures of which they cut up the Joynts that were served to His Majesty's Table.

Their houses are very ill built, the Walls bevil, without one right Angle in any Apartment; and this Defect ariseth from the Contempt they bear for practical Geometry; which they despise as vulgar and mechanick, those Instructions they give being too refined for the Intellectuals of their Workmen; which occasions perpetual Mistakes. And although they are dextrous enough upon a Piece of Paper, in the

Management of the Rule, the Pencil, and the Divider, yet in the common Actions and Behaviour of Life, I have not seen a more clumsy, awkward, and unhandy People, nor so slow and perplexed in their Conceptions upon all other Subjects, except those of Mathematicks and Musick. They are very bad Reasoners, and vehemently given to Opposition, unless when they happen to be of the right Opinion, which is seldom their Case. Imagination, Fancy, and Invention, they are wholly Strangers to, nor have any Words in their Language by which those Ideas can be expressed; the whole Compass of their thoughts and Mind, being shut up within the two forementioned Sciences.

Most of them, and especially those who deal in the Astronomical Part, have great Faith in judicial Astrology, although they are ashamed to own it publickly. But, what I chiefly admired, and thought altogether unaccountable, was the strong Disposition I observed in them towards News and Politicks; perpetually enquiring into publick Affairs, giving their Judgments in Matters of State; and passionately disputing every Inch of a Party Opinion. I have indeed observed the same Disposition among most of the Mathematicians I have known in *Europe;* although I could never discover the least Analogy between the two Sciences; unless those People suppose, that because the smallest Circle hath as many Degrees as the largest, therefore the Regulation and Management of the World require no more Abilities than the handling and turning of a Globe. But, I rather take this Quality to spring from a very common Infirmity of human Nature, inclining us to be more curious and conceited in Matters where we have least Concern, and for which we are least adapted either by Study or Nature....

I was at the Mathematical School, where the Master taught his Pupils after a Method scarce imaginable to us in *Europe*. The Proposition and Demonstration were fairly written on a thin Wafer, with Ink composed of a Cephalick Tincture. This the Student was to swallow upon a fasting Stomach, and for three Days following eat nothing but Bread and Water. As the Wafer digested, the Tincture mounted to his Brain, bearing the Proposition along with it. But the Success hath not hitherto been answerable, partly by some Error in the *Quantum* or Composition, and partly by the Perverseness of Lads; to whom this Bolus is so nauseous, that they generally steal aside, and discharge it upwards before it can operate; neither have they been yet persuaded to use so long an Abstinence as the Prescription requires.

Jonathan Swift,
"Voyage to Laputa," from
Gulliver's Travels, 1726

Achilles had overtaken the Tortoise, and had seated himself comfortably on its back.

"So you've got to the end of our race-course?" said the Tortoise. "Even though it *does* consist of an infinite series of distances? I thought some wiseacre or other [Zeno] had proved that the thing couldn't be done?"

"It *can* be done," said Achilles. "It *has* been done! *Solvitur ambulando.* You see the distances were constantly *diminishing:* and so—"

"But if they had been constantly *increasing*?" the Tortoise interrupted. "How then?"

"Then I shouldn't be *here*," Achilles modestly replied; "and *you* would have got several times round the world, by this time!"

"You flatter me—*flatten,* I mean," said the Tortoise; "for you *are* a heavy weight, and *no* mistake! Well now, would you like to hear of a race-course, that most people fancy they can get to the end of in two or three steps, while it *really* consists of an infinite number of distances, each one longer than the previous one?"

"Very much indeed!" said the Grecian warrior, as he drew from his helmet (few Grecian warriors possessed *pockets* in those days) an enormous note-book and a pencil. "Proceed! And speak *slowly,* please! *Shorthand* isn't invented yet!"

"That beautiful first proposition of Euclid!" the Tortoise murmured dreamily. "You admire Euclid?"

"Passionately! So far, at least, as one *can* admire a treatise that won't be published for some centuries to come!"

"Well, now, let's take a little bit of the argument in that first proposition—just *two* steps, and the conclusion drawn from them. Kindly enter them in your note-book. And, in order to refer to them conveniently, let's call them *A, B,* and *Z*:

(*A*) Things that are equal to the same are equal to each other.

(*B*) The two sides of this Triangle are things that are equal to the same.

(*Z*) The two sides of this Triangle are equal to each other.

"Readers of Euclid will grant, I suppose, that *Z* follows logically from *A* and *B,* so that any one who accepts *A* and *B* as true, *must* accept *Z* as true?"

"Undoubtedly! The youngest child in a High School—as soon as High Schools are invented, which will not be till some two thousand years later—will grant *that*."

"And if some reader had *not* yet accepted *A* and *B* as true, he might still accept the *Sequence* as a *valid* one, I suppose?"

"No doubt such a reader might exist. He might say 'I accept as true the Hypothetical proposition that, if A and B be true, Z must be true; but I *don't* accept A and B as true.' Such a reader would do wisely in abandoning Euclid, and taking to football."

"And might there not *also* be some reader who would say 'I accept A and B as true, but I *don't* accept the Hypothetical'?"

"Certainly there might. *He,* also, had better take to football."

"And *neither* of these readers," the Tortoise continued, "is *as yet* under any logical necessity to accept Z as true?"

"Quite so," Achilles assented.

"Well, now, I want you to consider *me* as a reader of the *second* kind, and to force me, logically, to accept Z as true."

"A tortoise playing football would be—" Achilles was beginning.

"—an anomaly, of course," the Tortoise hastily interrupted. "Don't wander from the point. Let's have Z first, and football afterwards!"

"I'm to force you to accept Z, am I?" Achilles said musingly. "And your present position is that you accept A and B, but you *don't* accept the Hypothetical—"

"Let's call it C," said the Tortoise.

"—but you don't accept:

(C) If A and B are true, Z must be true."

"That is my present position," said the Tortoise.

"Then I must ask you to accept C."

"I'll do so," said the Tortoise, "as soon as you've entered it in that note-book of yours. What else have you got in it?"

"Only a few memoranda," said Achilles, nervously fluttering the leaves: "a few memoranda of—of the battles in which I have distinguished myself!"

"Plenty of blank leaves, I see!" the Tortoise cheerily remarked. "We shall need them *all!*" (Achilles shuddered.) "Now write as I dictate:

(A) Things that are equal to the same are equal to each other.

(B) The two sides of this Triangle are things that are equal to the same.

(C) If A and B are true, Z must be true.

(Z) The two sides of this Triangle are equal to each other."

"You should call it D, not Z," said Achilles. "It comes *next* to the other three. If you accept A and B *and* C you *must* accept Z."

"And why *must* I?"

"Because it follows *logically* from them. If A and B and C are true, Z *must* be true. You don't dispute *that,* I imagine?"

"If A and B and C are true, Z *must* be true," the Tortoise thoughtfully repeated. "That's *another* Hypothetical, isn't it? And, if I failed to see its truth, I might accept A and B and C, and *still* not accept Z, mightn't I?"

"You might," the candid hero admitted; "though such obtuseness would certainly be phenomenal. Still, the event is *possible.* So I must ask you to grant one more Hypothetical."

"Very good. I'm quite willing to grant it, as soon as you've written it down. We will call it

(D) If A and B and C are true, Z must be true.

"Have you entered that in your note-book?"

"I *have!*" Achilles joyfully exclaimed, as he ran the pencil into its sheath. "And at last we've got to the end of this ideal race-course! Now that you accept A and B and C and D, *of course* you accept Z."

"Do I?" said the Tortoise innocently. "Let's make that quite clear. I accept A

and *B* and *C* and *D*. Suppose I *still* refuse to accept *Z*."

"Then Logic would take you by the throat, and *force* you to do it!" Achilles triumphantly replied. "Logic would tell you 'You can't help yourself. Now that you've accepted *A* and *B* and *C* and *D*, you *must* accept *Z*!' So, you've no choice, you see."

"Whatever *Logic* is good enough to tell me is worth *writing down*," said the Tortoise. "So enter it in your book, please. We will call it

(*E*) If *A* and *B* and *C* and *D* are true, *Z* must be true.

"Until I've granted *that*, of course, I needn't grant *Z*. So it's quite a *necessary* step, you see?"

"I see," said Achilles; and there was a touch of sadness in his tone.

Here the narrator, having pressing business at the Bank, was obliged to leave the happy pair, and did not again pass the spot until some months afterwards. When he did so, Achilles was still seated on the back of the much-enduring Tortoise, and was writing in his note-book, which appeared to be nearly full. The Tortoise was saying, "Have you got that last step written down? Unless I've lost count, that makes a thousand and one. There are several millions more to come. And *would* you mind, as a personal favour—considering what a lot of instruction this colloquy of ours will provide for the Logicians of the Nineteenth Century—*would* you mind adopting a pun that my cousin the Mock-Turtle will then make, and allowing yourself to be re-named Taught-Us?"

"As you please!" replied the weary warrior, in the hollow tones of despair, as he buried his face in his hands. "Provided that *you*, for *your* part, will adopt a pun the Mock-Turtle never

made, and allow yourself to be renamed A Kill-Ease!"

Lewis Carroll
[Charles L. Dodgson],
"What the Tortoise Said
to Achilles," 1894

In the night I had a dream.

I saw before me a vast multitude of small Straight Lines (which I naturally assumed to be Women) interspersed with other Beings still smaller and of the nature of lustrous points—all moving to and fro in one and the same Straight Line, and, as nearly as I could judge, with the same velocity.

A noise of confused, multitudinous chirping or twittering issued from them at intervals as long as they were moving; but sometimes they ceased from motion, and all was silence.

Approaching one of the largest of what I thought to be Women, I accosted her, but received no answer. A second

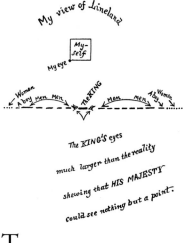

The strange, mathematical world of Lineland.

and a third appeal on my part were equally ineffectual. Losing patience at what appeared to me intolerable rudeness, I brought my mouth into a position full in front of her mouth so as to intercept her motion, and loudly repeated my question, "Woman, what signifies this concourse, and this strange and confused chirping, and this monotonous motion to and fro in one and the same Straight Line?"

"I am no Woman," replied the small Line: "I am the Monarch of the world. But thou, whence intrudest thou into my realm of Lineland?" Receiving this abrupt reply, I begged pardon if I had in any way startled or molested his Royal Highness; and describing myself as a stranger I besought the King to give me some account of his dominions....

It seemed that this poor ignorant Monarch—as he called himself—was persuaded that the Straight Line which he called his Kingdom, and in which he passed his existence, constituted the whole of the world, and indeed the whole of Space. Not being able either to move or to see, save in his Straight Line, he had no conception of anything out of it. Though he had heard my voice when I first addressed him, the sounds had come to him in a manner so contrary to his experience that he had made no answer, "seeing no man," as he expressed it, "and hearing a voice as it were from my own intestines." Until the moment when I placed my mouth in his World, he had neither seen me, nor heard anything except confused sounds beating against—what I called his side, but what he called his *inside* or *stomach;* nor had he even now the least conception of the region from which I had come. Outside his World, or Line, all was a blank to him; nay, not even a blank, for a blank implies Space; say, rather, all was non-existent.

His subjects—of whom the small Lines were men and the Points Women—were all alike confined in motion and eyesight to that single Straight Line, which was their World. It need scarcely be added that the whole of their horizon was limited to a Point; nor could any one ever see anything but a Point. Man, woman, child, thing—each was a Point to the eye of a Linelander. Only by the sound of the voice could sex or age be distinguished. Moreover, as each individual occupied the whole of the narrow path, so to speak, which constituted his Universe, and no one could move to the right or left to make way for passers by, it followed that no Linelander could ever pass another. Once neighbours, always neighbours. Neighbourhood with them was like marriage with us. Neighbours remained neighbours till death did them part.

Such a life, with all vision limited to a Point, and all motion to a Straight Line, seemed to me inexpressibly dreary; and I was surprised to note the vivacity and cheerfulness of the King. Wondering whether it was possible, amid circumstances so unfavourable to domestic relations, to enjoy the pleasures of conjugal union, I hesitated for some time to question his Royal Highness on so delicate a subject; but at last I plunged into it by abruptly inquiring as to the health of his family. "My wives and children," he replied, "are well and happy."

Staggered at this answer—for in the immediate proximity of the Monarch (as I had noted in my dream before I entered Lineland) there were none but Men—I ventured to reply, "Pardon me, but I cannot imagine how your Royal Highness can at any time either see or approach their Majesties, when there are at least half a dozen intervening individuals, whom you can neither see through, nor pass by? Is it possible that in Lineland proximity is not necessary for marriage and for the generation of children?"

"How can you ask so absurd a question?" replied the Monarch. "If it were indeed as you suggest, the Universe would soon be depopulated. No, no; neighbourhood is needless for the union of hearts; and the birth of children is too important a matter to have been allowed to depend upon such an accident as proximity. You cannot be ignorant of this. Yet since you are pleased to affect ignorance, I will instruct you as if you were the veriest baby in Lineland. Know, then, that marriages are consummated by means of the faculty of sound and the sense of hearing.

"You are of course aware that every Man has two mouths or voices—as well as two eyes—a bass at one and a tenor at the other of his extremities. I should not mention this, but that I have been unable to distinguish your tenor in the course of our conversation." I replied that I had but one voice, and that I had not been aware that his Royal Highness had two. "That confirms my impression," said the King, "that you are not a Man, but a feminine Monstrosity with a bass voice, and an utterly uneducated ear. But to continue.

"Nature having herself ordained that every Man should wed two wives—" "Why two?" asked I. "You carry your affected simplicity too far," he cried. "How can there be a completely harmonious union without the combination of the Four in One, viz. The Bass and Tenor of the Man and the Soprano and Contralto of the two Women?" "But supposing," said I, "that a man should prefer one wife or three?" "It is impossible," he said; "it is as

inconceivable as that two and one should make five, or that the human eye should see a Straight Line."

Edwin A. Abbott [A. Square],
"How I Had a Vison of Lineland,"
from *Flatland: A Romance of Many Dimensions,* 1884

"Are there any precious stones in [the mine]?" asked Milo excitedly.

"…I'll say there are. Look here." [The Mathemagician] reached into one of the carts and pulled out a small object, which…sparkled brightly.

"But that's a five," objected Milo, for that was certainly what it was.

"Exactly," agreed the Mathemagician; "as valuable a jewel as you'll find anywhere. Look at some of the others."

He scooped up a great handful of stones and poured them into Milo's arms. They included all the numbers from one to nine, and even an assortment of zeros.

"We dig them and polish them right here," volunteered the Dodecahedron, pointing to a group of workers busily employed at the buffing wheels; "and then we send them all over the world. Marvelous, aren't they?"

"They are exceptional," said Tock, who had a special fondness for numbers.

"So that's where they come from," said Milo, looking in awe at the glittering collection of numbers. He returned them to the Dodecahedron… but, as he did, one dropped to the floor with a smash and broke in two.…

"Oh, don't worry about that," said the Mathemagician as he scooped up the pieces. "We use the broken ones for fractions."

Norton Juster,
The Phantom Tollbooth, 1961

…And an amusing puzzle:

Cat and mice

Purrer has decided to take a nap. He dreams he is encircled by 13 mice: 12 gray and 1 white. He hears his owner saying: "Purrer, you are to eat each thirteenth mouse, keeping the same direction. The last mouse you eat must be the white one." Which mouse should he start from?

Purrer the cat dreams of the mathematically correct way to eat mice.

Boris A. Kordemsky,
*The Moscow Puzzles:
359 Mathematical Recreations,* 1956,
trans. Albert Parry, 1972

Solution

Start from the cross in the diagram (position 13) and go clockwise through positions 1, 2, 3,…, crossing out each thirteenth dot: 13, 1, 3, 6, 10, 5, 2, 4, 9, 11, 12, 7, and 8. Call position 8 the white mouse, and Purrer starts clockwise from the fifth mouse clockwise from the white mouse (i.e., position 13 relative to position 8). Or he starts counterclockwise from the fifth mouse counterclockwise from the white mouse.

Glossary

Principal sources for definitions are Parker and Gibson (see **Further Reading**) *and* Webster's Third New International Dictionary of the English Language.

Abacus A calculating device in which rows of beads strung on parallel wires or strings are set in a frame. Each row represents one level of place value; each bead represents a digit.

Abundant number A positive integer that is greater than the sum of its own divisors.

Additive numeration system A numeration system in which addition is the only operation employed in the formation of numbers. A number is created by the juxtaposition of symbols; its value is equal to the sum of the values of the symbols. Each cluster of symbols is repeated as many times as required; thus, *two hundred* is conceptualized as *one hundred + one hundred,* and is represented by the repetition two times of the digit representing *one hundred.* The Roman numeral system is additive.

Aleph A letter of the Hebrew alphabet used to represent a transfinite number.

Algebra A branch of mathematics in which arithmetic problems and operations are expressed in symbols that represent numbers and variables.

Algorism A medieval term for computation according to the Indo-Arabic method; an early form of algebra.

Alphabetical numeration system A numeration system in which digits do not have their own symbols but adopt letters of the alphabet, each assigned a numerical value; the Classical Greek number system is alphabetical.

Amicable numbers A pair of numbers, each of which is equal to the sum of all the exact divisors of the other except the number itself, e.g., 220 and 284.

Apices In an abacus or computing table, the elements or objects that represent individual digits and are used for counting.

Arabic numerals In modern usage, the numerals employed in the West and much of the world today: 1, 2, 3, 4, 5, 6, 7, 8, 9, 0. Sometimes called *Indo-Arabic* or *Hindu-Arabic numerals,* these figures derive from North African Arabic figures, called *Gobar numerals,* whose early source is Hindu India.

Arithmetic The science of numbers; the body of knowledge, functions, and principles used for the study of calculation and the solving of numerical problems. Basic operations of arithmetic include addition, subtraction, division, multiplication, raising to a power, and extracting a root.

Base A defining characteristic in a number system that determines place value. E.g., in a decimal system, the base is ten, and place value is established in powers of ten: ones, tenths, hundredths, etc.; in a binary system, the base is two.

Base system A system that permits efficient counting and management of numbers in bundles; see also **Place value.**

Binary notation A means of representing numbers using only the digits 0 and 1.

Calculation The act or process of carrying out a mathematical operation that includes two or more sets of numbers; reckoning with numbers.

Calculus A branch of mathematics that provides formulas and rules for the calculation of irregular or changing quantities and measures, such as rates of change, speed, or motion, and semiregular or irregular volumes and areas, such as curves and cones.

Cardinal numbers The integers as used for counting and to denote number or quantity: e.g., one, two, three, etc.

Complex number A number that has a compound form in which one part is a real number and the other an imaginary number (a multiple of the number $\sqrt{-1}$, notated as i). A complex number takes the form $a + ib$, a and b being real numbers. The set of complex numbers is notated as C.

Complex plane A plane, represented in graph form, whose horizontal axis is the real part of a complex number and whose vertical axis is its imaginary part; also called an *Argand diagram.*

Conjecture A mathematical statement whose truth is proposed but not proved.

Connectivity The theory that smaller numbers exist between any two given numbers, and that these numbers can be named, calculated, and expressed mathematically.

Constant In an equation, a quantity whose value is unchanging, usually defined in relation to a variable.

Continuum A concept of numerical infinity, expressed by the real number line, an imaginary line running to infinity in both directions from a point 0, along which numbers lie with infinite density. For every two points on the continuum there is always an infinite number of points between the two.

Cube In algebra, a number that is the threefold product of its own quantity: $a \times a \times a$; notated as a^3.

Decimal A number system organized on base ten; a decimal fraction is a number smaller than one, expressed in the decimal system, that is, in terms of tenths, hundredths, thousandths, etc.

Deficient number A positive integer that is less than the sum of its own divisors.

Denominator The lower number a fraction (below the line expressing division) that divides the upper number; the divisor.

Denumerable set A set of numbers that can be counted, even if that set is infinite; a set of numbers that has a one-to-one correspondence with the set of positive integers.

Difference The result of the operation of subtraction.

Digit A written sign or character that represents a single element of a number; a numeral used in a position within a number. The term *digit*, as opposed to *numeral*, implies place value.

Digital calculation The use of the fingers (and sometimes other parts of the body) as a mnemonic tool in counting and computing.

Divisor In the operation of division, the number or quantity by which another number is divided; e.g., in the operation $18 \div 2 = 9$, 2 is the divisor.

Equation A statement of equality between two mathematical expressions, such as numbers, functions, magnitudes, and operations.

Even number A natural number divisible by 2.

Exponent A number or symbol placed in the superscript position to the right of a number or expression that indicates its power.

Expression A combination of mathematical symbols (which may include numbers) and operations.

Factor A divisor; the number or quantity by which another number is divided.

Figural numeration A concrete, orderly, written notation system for numbers, composed of written marks, each of which is assigned a numerical identity.

Figurate numbers A Pythagorean concept in which numbers and sequences of numbers are linked to the specific geometric figures they engender; e.g., triangular numbers include 3, 6, 10, and 15; polygonal numbers also exist (oblong, square, hexagonal, etc.), and polyhedral numbers, all of which are assigned spiritual meanings and values as well as purely mathematical ones.

Finite A number or equation that has definite or definable limits or boundaries. A finite set is one that has a fixed, countable number of elements.

Fraction Part of a natural number, expressed as a number (the denominator) divided by another number (the numerator). A fraction is written as the numerator over the denominator, divided by a horizontal line that represents the operation of division.

Giga– A prefix denoting *billion* (10^9).

Golden number A number that represents the ratio $\frac{1}{a} = \frac{a}{b}$, expressed mathematically as $\frac{1+\sqrt{5}}{2}$. A proportion of particular aesthetic elegance, much used in Classical art and architecture, and in golden section (line) and golden rectangle geometry, it is represented by the figure Φ.

Hybrid numeration system A numeration system that exploits both addition and multiplication. Addition is used as in additive numeration to count up the contributions of successive powers. But within each power multiplication is used in the following way: *two hundred* is conceptualized as *two* times *hundred* and is represented by *two* followed by *hundred*. This juxtaposition is, in effect, a multiplication.

Hypotenuse The side of a right-angled triangle that is opposite to the right angle.

Imaginary axis The vertical axis of a complex plane, or Argand diagram, used for the expression of complex numbers. It represents the part of a complex number that is imaginary (a multiple of $\sqrt{-1}$).

Imaginary number The entity, not a real number, whose square is equal to –1; notated as *i*.

Incommensurability Lacking a common basis for comparison; in number theory, two numbers are said to be incommensurable when their ratio is irrational.

Indo-Arabic numerals See **Arabic numerals**.

Infinite A quantity, value, or measure larger than any fixed number. An infinite set is one that contains a defined collection of elements whose number is infinite; a set that can be placed in one-to-one correspondence with a subset of itself.

Infinite decimal A decimal that in expressing a fraction does not end; an infinite decimal that is a rational number is a repeating decimal (e.g., 0.55555…); an irrational number, such as π, can be expressed as a nonrepeating infinite decimal.

Infinitesimal A quantity, value, or measure smaller than any fixed number, but not equal to zero.

Infinity Not a number, but the concept of a value larger than any measured or defined value; the value of a quantity that increases without limit; expressed by the symbol ∞. Modern mathematics distinguishes among kinds of infinities, or infinities of differing order or degree.

Integer A number belonging to the set of natural numbers, used for counting, their negatives, and zero; notated as $\mathbb{Z} = \{\ldots, -2, -1, 0, 1, 2,\ldots\}$.

Irrational number A real number that cannot be written as a fraction or expressed as the quotient of two integers.

Magnitude The numerical value or measure assigned to a mathematical or physical quantity, by which it may be compared with other quantities.

Mersenne number A kind of prime number that can be expressed as a power of two minus one, e.g., 3, 7, and 31 are all Mersenne primes: $2^2 - 1 = 3$, $2^3 - 1 = 7$, $2^5 - 1 = 31$. The largest known prime numbers are all Mersenne numbers.

Milli– A prefix denoting *thousandth* (10^{-3}).

Natural number A positive integer; a whole number. The set of natural numbers, notated as $\mathbb{N} = \{0, 1, 2, 3,\ldots\}$, is obtained by adding 1 to 0 and then adding 1 to that number and repeating the process infinitely.

Natural sequence A concept, basic to the idea of number, that a group of numbers can be placed in a sequence following a logical pattern or progression.

Nondenumerable set A set of numbers that cannot be counted; one that cannot be placed in one-to-one correspondence with the set of positive integers.

Notation The use of symbols to represent or express quantities and operations.

Notation system See **Numeration system**.

Notational digit A natural number as used in positional notation; that is, to indicate quantity by numerical value and place value together. (In base 10 the natural numbers from 1 to 9 plus 0 are used.)

Numeral A single written sign or character that represents a number; also, a word responsible for verbalizing a number.

Numerator The upper number in a fraction (above the line expressing division) that is divided by the lower number; the dividend.

Numeration, numeration system A means of naming numbers; a method of representing numbers by numerals, in which each numeral represents only one number. Such systems may be spoken or written, additive or hybrid, etc. A numeration system formalizes numbers, making them concrete and perceptible, rather than abstract and ephemeral.

Odd number A natural number not divisible by 2.

One-to-one correspondence A method of defining or distinguishing finite from infinite and denumerable from nondenumerable sets or classes of elements (such as numbers) by the pairing of two sets, in which each element of each set is made to correspond to one and only one element of the other set.

Operation A defined mathematical procedure or activity between two numbers, such as addition, subtraction, multiplication, division, and aspects of these, such as raising to a power or extracting a root.

Ordinal numbers The integers as used to denote order or rank, rather than number or quantity: e.g., first, second, third, etc.

Perfect number A positive integer equal to the sum of its own factors.

Pi The ratio of the circumference of any circle to its diameter, an irrational number, notated as π.

Place value The value given to a digit by virtue of its location within a number; also called *position*.

Positional notation A numeration system, usually hybrid, in which a number is represented by a group of digits each of whose value is indicated by a combination of its place in the sequence and its numerical value; also called *positional notation*.

Positive integer See **Natural number.**

Power The number of times a quantity or number is multiplied by itself, expressed as a number and an exponent. The repetition of multiplications of a number or quantity is called *raising to a power.*

Prime number A positive integer that is divisible only by 1 and by itself.

Prime pair A pair of prime numbers that differ by 2; e.g., 17 and 19.

Product The result of the operation of multiplication.

Quipu An Inca calculating and record-keeping device used by the Inca, consisting of a group of strings with knots in them, attached to a main cord. The position of each knot on a string, the lengths and number of strings, their position on the cord, and colors and types of knot and string all convey numerical information.

Quotient The result of the operation of division.

Rational number An integer or fraction, including all finite and repeating decimals; the set of rational numbers is notated as Q.

Real axis The horizontal axis of a complex plane, used for the expression of complex numbers. It represents the part of a complex number that is real.

Real number Any rational or irrational number; the set of real numbers is notated as R.

Real number line A graphic representation of all numbers in the form of an infinite straight line, marked with a point 0, to the left of which lie negative numbers and to the right positives. This model describes the position of all numbers relative to 0 and to each other, and expresses the concept of continuity; see also **Continuum.**

Repeating decimal See **Infinite decimal.**

Set Any group of things or numbers that belong to a defined category. In mathematics, a group like numbers forms a set. A finite set has a limited and countable number of elements; an infinite set has a defined but unlimited number of elements. A set is notated thus: $X = \{a, b, c, d, e,...\}$.

Set theory The study of the structure, size, and properties of sets.

Sexagesimal A numeration system based on the multiples of 60.

Singleton A set with one element.

Spoken numeration The process of giving names to numbers, or of verbalizing concepts of number.

Square In algebra, a number that is the twofold product of its own quantity, $a \times a$; notated as a number and its exponent: a^2.

Square root A factor of a number that when squared yields the number, e.g., either +4 or −4 is the square root of 16.

Sum The result of the operation of addition.

Transcendental number A complex number that is not algebraic and cannot be written as an exact ratio of two integers; a kind of irrational number.

Transfinite number A number assigned to an infinite set, indicating the power, or degree, of its denumerability or its infinity.

Trigonometry The study of the relationships among the sides and angles of triangles and related magnitudes.

Unit The numeral 1; the first natural number.

Variable In an equation, a changing or undetermined quantity with some known characteristics or a range of possible values; usually indicated by a letter and often defined in relation to a constant.

Vector A measure having both quantity or magnitude and direction.

Whole numbers The positive integers; the numbers used for counting; the natural numbers.

Written numeration A means of representing numbers and concepts of number using no words but only symbols (digits, numerals).

Zero The symbol for a quantity of none; the number that when added to another number does not increase its value.

Chronology

*The dating of inventions and progress in the history of mathematics is not always precise. Many early discoveries occurred in several cultures at different times. Principal sources for this time line are Motz and Boyer and Merzbach (see **Further Reading**).*

BC

c. 30,000: Paleolithic tally bones with numerical notches

4th millennium: appearance of *calculi,* clay counting stones, in Mesopotamia and other regions of the Middle East

c. 3300: writing, and the first digits, invented in Sumer and Elam

3d millennium: use of hieroglyphic numerals and additive, decimal notation in Egypt

c. 2700: Sumerian cuneiform digits in use

c. 2000: appearance of the decimal base system

c. 1800: positional notation appears in Babylon

c. 1700: the Rhind Papyrus, in Egypt, in which the scribe Ahmes explores the measurement of the area of a circle

c. 1450: earliest Chinese notation system

c. 1300: appearance of digits in China

6th century: Pythagoras (c. 580–c. 500), Greek mathematical philosopher, founds the Pythagorean school in southern Italy; Pythagoreans distinguish between odd and even numbers; develop concept of connectivity and other theories of number

5th century: Greek philosophers who advance concepts of mathematics include Philolaus of Croton (470–390), Anaxagoras of Clazomenae (c. 500–c. 428), Zeno of Elea (c. 495–c. 430), Democritus of Abdera (c. 460–370), Plato (c. 428–348), in Athens; in Persia, knotted cords used for calculation

c. 425: the discovery of incommensurable magnitudes and irrational values disrupts Pythagorean mathematics; the discovery is sometimes credited to Hippasus of Metapontum, a former Pythagorean

4th century: in Greece, alphabetic numeration system in use; Aristotle (384–322) develops first significant mathematical concept of infinity

c. 300: Euclid, in Alexandria, writes the *Elements,* establishing mathematical method of postulate and proof; foundation of Alexandria as a center of mathematics study

3d century: appearance of the first zero, in Babylon; numerals partially resembling modern Indo-Arabic numerals appear in inscriptions in India; Archimedes (c. 287–212), Greek inventor and mathematician, writes a treatise on π

2d century: use in China of positional numeration, without the zero

c. 140: Hipparchus, Greek astronomer, invents an early form of trigonometry

1st century: Chinese place-value system

AD

c. 1st century: first use of negative numbers

2d century: paper invented in China

120: in China, Chang Hing calculates π as 3.1555…

4th–5th century: Indian positional notation, with zero

415: the death of Hypatia (c. 370–415), Alexandrian Greek philosopher and author of texts on mathematics, marks the close of the golden age of Alexandrian mathematics

458: the *Lokavibhaga (The Parts of the Universe),* a Sanskrit treatise, uses the place-value system

5th–9th century: the Maya place-value system, with zero, in use in Mesoamerica

6th century: Boethius (c. 480–524), Roman philosopher and mathematician, writes treatises on logic and arithmetic; in India, Āryabhata I (476–c. 550) refines the calculation of π to 3.1416

8th century: in England, Bede (c. 673–735) writes a treatise on digital calculation; Indian calculation method reaches Baghdad: beginning of the golden age of Arabic mathematics

9th century: Thābit ibn Qurrah (c. 836–901), Arabic mathematician, describes amicable numbers

c. 825: in Baghdad, Muhammad ibn Mūsā al-Khwārizmī (c. 780–c. 850), called the "father of algebra," writes the *Kitab al-jabr wa al-muqa balah (Treatise on Restoration or Completion and of Reduction or Balancing),* on algebra

10th century: the mathematician and poet Omar Khayyám (1048?–1131), in Persia, develops a general number theory and publishes an influential treatise on algebra; Gobar numerals in use in North Africa and the Iberian Peninsula; these differ from the Indo-Arabic numerals in use in the Arab Middle East and are the direct ancestors of the numerals used today in the West

12th century: in India, the *Bakshali* manuscript on arithmetic is written, although parts of it may be much older; in Europe, Arabic numerals (without zero) begin to appear; they are definitively in use there by the 14th century

13th century: Sacrobosco (John of Halifax, c. 1200–1256), English mathematician, writes *General Algorism;* in Peru, Inca *quipu,* knotted calculating cord, in use; in Mongolia, Nasir Eddin al-Tusi (1201–74), writes on trigonometry and astronomy and offers a proof of the parallel postulate; decline of the Arab golden age of mathematics

c. 1200: in France Alexandre de Villedieu writes the *Carmen de algorismo (Poem on Algorism),* describing the operations of integers

1202: Fibonacci (Leonardo of Pisa, c. 1180 1250) writes the *Liber abaci (Book of the Abacus),* on algebra

14th century: the zero appears in Europe

15th century: with the development of the printing

press, Indo-Arabic numerals acquire a definitive form in Europe; negative numbers appear

1427: al-Kashi (died c. 1436), in Samarkand, defines decimal fractions and writes the *Miftah al-hisab (The Key to Arithmetic)*

c. 1450–92: Piero della Francesca (1410?–92), Italian painter and mathematician, writes the treatises *Trattato d'abaco (Abacus Treatise)* and *De prospectiva pingendi (On Perspective for Painting)*, and an essay on Euclid, uniting mathematics and aesthetic theory

1484: Nicolas Chuquet (c. 1445–c. 1500), in France, writes the *Triparty en la science des nombres (Three Parts in the Science of Numbers)*, on fractions, decimals, and irrational numbers

1489: Johannes Widman (c. 1462–98?), in Germany, first uses the plus and minus signs (+ and –) in print

1492: in Italy, Francesco Pellos (c. 1450–c. 1500) first uses the decimal point, in his *Compendio de lo abaco (Compendium of the Abacus)*

1494: Luca Pacioli (1445?–1514?), in Italy, writes the *Summa de arithmetica, geometrica, proportioni et proportionalità (Summary of Arithmetic, Geometry, Proportions, and Proportionality)*

16th century: galley division in use in Europe; in China a form of digital calculation permits computations involving numbers in the billions

1545: Geronimo Cardano (1501–76), in Italy, publishes the *Ars magna (Great Art)*, giving the solutions of cubic and quartic equations, and using a negative square root; with Rafael Bombelli (c. 1526–73) he describes complex numbers for the first time

1556–60: Niccolò Tartaglia (1499–1557), in Italy, publishes his *Trattato di numeri et misure (Treatise on Numbers and Measures)*

1557: in England, Robert Recorde (1501–58) writes a treatise on algebra, *The Whetstone of Witte*, in which the sign for equality (=) first appears

1585: Simon Stevin (1548–1620), in Holland, writes *De thiende (The Tenth)*, on decimals

1614: John Napier (1550–1617), a Scot, invents logarithms

1632: Galileo Galilei (1564–1642) Italian mathematician and astronomer, writes *The Two Chief Systems* and, in 1638, *The Two New Sciences*, texts that explore the infinite and the infinitesimal

1635: Francesco Cavalieri (1598–1647), in Italy, originates the method of indivisibles, precursor of integral calculus; he also writes on trigonometry

1637: René Descartes (1596–1650), in France and Holland, publishes the *Discours de la méthode (Discourse on Method)*, on establishing mathematical certainty

1640: Pierre de Fermat (1601–65), in France, formulates his famous "last conjecture," stating that there is no cube divisible into two cubes; he contributes to the foundation of modern number theory and differential calculus

c. 1642: Blaise Pascal (1632–62), in France, invents a Calculating Machine, forerunner of the computer, and contributes to number theory

1644: Marin Mersenne (1588–1648), in France, advances the study of prime numbers

c. 1650: John Wallis (1616–1703), in England, researches the problem of π

1665: Isaac Newton (1642–1727), English physicist and mathematician, writes on the calculus; in 1687, publishes *Philosophiae naturalis principia mathematica* (known as the *Principia*)

1684: Gottfried Wilhelm Leibniz (1646–1716), in Germany, publishes his first paper on the differential calculus; in 1686 he describes the integral calculus; in 1703 he first uses the binary system

c. 1700: the ratio of a circle's diameter to its circumference is named *pi*, written π

1742: Christian Goldbach (1690–1764), in Germany and Russia, writes on prime numbers, contributes to the study of number theory, and formulates the "Goldbach conjecture," stating that every even natural number is the sum of two prime numbers

1754: Jean-Etienne Montucla (1725–99), in France, writes the *Histoire des recherches sur la quadrature du cercle (History of Inquiries on Squaring the Circle)* and, in 1758, *Histoire des mathématiques (History of Mathematics)*

1761: Johann Heinrich Lambert (1728–77), in Germany, offers the first proof that π is an irrational number

c. 1777: Leonhard Euler (1707–83), Swiss mathematician in Russia and Germany, devises the modern notation system for calculus, with first use of *i* to signify the square root of –1, an imaginary number; he pursues research on integral calculus and develops theories of complex numbers

1797: Caspar Wessel (1745–1818), in Norway, develops the idea of the complex plane

1801: Carl Friedrich Gauss (1777–1855), in Germany, proposes the method of least squares and a solution for binomial equations

1806: Jean Robert Argand (1768–1822), in Switzerland, devises the Argand diagram for complex numbers

1830: Evariste Galois (1811–32), in France, researches algebraic numbers, equations, and equation theory

1833: William Rowan Hamilton (1805–65), in Ireland, introduces a formal algebra of real number

1834: Nicolai Ivanovich Lobachevsky (1792–1856), a Russian, discovers a method of approximating the roots of algebraic equations

1844: Joseph Liouville (1809–82), in France, identifies a class of nonalgebraic real numbers, or transcendental numbers, now called Liouville numbers

1872: Richard Dedekind (1831–1916), in Germany, develops (with Georg Cantor) an arithmetic theory of irrational numbers, publishing *Stetigkeit und irrationale Zahlen (Continuity and Irrational Numbers)*

1874: Georg Cantor (1845–1918), in Germany, publishes *Mengenlehre (Set Theory)*, establishing modern set theory; he develops (with Richard Dedekind) theories of irrational numbers and of the arithmetic of actual infinity, introduces transfinite numbers, and explores new aspects of infinite numbers

1882: Ferdinand von Lindemann (1852–1939), in Germany, establishes the impossibility of squaring the circle, proving that π is a transcendental number

1893: David Hilbert (1862–1943), in Germany, publishes the *Zahlbericht (The Theory of Algebraic Number Fields,* literally, *Report on Numbers),* on number theory; he develops innovative theories of invariants and integral equations. Gottlob Frege (1848–1925), in Germany, writes *Grundgesetze der Arithmetik (The Basic Laws of Arithmetic)* and works on mathematical logic

1895: Henri Poincaré (1854–1912), in France, publishes *Analysis situs (Topology),* introducing the new mathematical field of topology

1896: proof of the prime number theorem, by Jacques

Hadamard (1865–1963), a Frenchman, and C. J. de la Vallée-Poussin (1866–1962), a Belgian, working independently

1903: Bertrand Russell (1872–1970), English mathematician and philosopher, publishes *Principles of Mathematics* and works on mathematical logic

1910–13: Russell and Alfred North Whitehead publish their *Principia mathematica (Principles of Mathematics)* in England

1931: Kurt Gödel (1906–78) an Austrian, proposes a theorem of undecidable propositions, demonstrating that within a logical arithmetical system some axioms can neither be proved nor disproved

1995: Andrew Wiles completes his proof of Fermat's conjecture (now theorem), in England

Further Reading

Badiou, A., *Le Nombre et les nombres,* 1990

Ball, W. W. Rouse, *A Short Account of the History of Mathematics,* 1960

Beaujouan, G., *Par raison de nombres: L'Art du calcul et les savoirs scientifiques médiévaux,* 1991

Boyer, C. B., and Merzbach, U. C., *A History of Mathematics,* 2d ed., 1991

Burton, D., *Elementary Number Theory,* 1980

Courant, R., and Robbins, H., *What Is Mathematics?,* 1969

Crump, T., *The Anthropology of Numbers,* 1990

Dantzig, T., *Number: The Language of Science,* 4th ed., 1954

Devlin, K., *Mathematics: The Science of Patterns,* 1994, 1997

Emmer, M., ed., *The Visual Mind: Art and Mathematics,* 1993

Eves, H., *An Introduction to the History of Mathematics,* 5th ed., 1983

Frege, G., *The Basic Laws of Arithmetic,* 1982

Gibson, C., ed., *The Facts on File Dictionary of Mathematics,* 1981

Guitel, G., *Histoire comparée des numérations écrites,* 1975

Hogben, L., *The Wonderful World of Mathematics,* 1955

Ifrah, G., *From One to Zero: A Universal History of Numbers,* 1985

Lawlor, R., *Sacred Geometry: Philosophy and Practice,* 1982

Levy, T., *Figures de l'infini: Les mathématiques au miroir des cultures,* 1987

Maor, E., *To Infinity and Beyond: A Cultural History of the Infinite,* 1987

McLeish, J., *Number: The History of Numbers and How They Shape Our Lives,* 1991

Motz, L., and Waver, J. H., *The Story of Mathematics,* 1993

Parker, S. P., ed., *McGraw-Hill Dictionary of Mathematics,* 1994

Péter, R., *Playing with Infinity: Mathematical Explorations and Excursions,* 1957, 1961

Phillips, R., *Numbers: Facts, Figures and Fiction,* 1994

Piaget, J., and A. Szeminska, *The Child's Conception of Number,* 1965

Pise, L. D., *Le Livre des nombres carrés,* 1952

Rashid, R., *The Development of Arabic Mathematics,* 1994

Stewart, I., *The Problems of Mathematics,* 1992

Wells, D., *The Penguin Dictionary of Curious and Interesting Numbers,* 1987

List of Illustrations

Index

Photograph Credits

© Harry N. Abrams, Inc., New York: front cover and spine. Harry N. Abrams, Inc. archives: 145a, 146. © ADAGP, Paris, 1996: back cover, 14–15, 62, 66a–67a, 86–87, 91, 96, 102, 104, 118, 119. Agence Vu, Paris, Arnaud Legrai: 64. AKG, Paris: 117. All rights reserved: 13, 31a, 36, 42–43, 49, 50c–51c, 56c, 66b, 67b, 70b, 71, 80c, 81, 86–87, 101, 103, 108c, 111, 113b, 120, 127, 142, 145b. Dieter Appelt, Berlin: 19a,b. Artephot, Paris, Oronoz: 25/Fabbri: 27. Barbe, Paris: 121. Bibliothèque Municipale, Toulouse, France: 76–77. Bibliothèque Nationale de France, Paris: 17, 18, 37, 52c–53, 72, 83, 84a, 89, 129, 137. Bodleian Library, Oxford: 38, 50b, 142. B.P.K., Berlin: 22–23. British Library, London: 54–55. British Museum, London: 54, 82c, 82b, 130. Bulloz, Paris: 125. Charmet, Paris: 102, 116. Miguel Chevalier, Paris: 58–59, 109, 122. Columbia University, New York: 46, 47. © Cordon Art, 1996, Baarn: 114–15. Cosmos, Paris S.P.L./Prof. P. Goddard: 77a. Dagli Orti, Paris: 12, 30–31, 34–35, 43, 92–93, 112–13a. Deutsches Museum, Munich, Germany: 147. Dover Publications, Inc.: 161, 164. Flammarion-Giraudon, Vanves, France: 58a, 143. Fotogram-Stone Images, Paris: 115b. Gallimard Archives: 28b, 40–41, 135, 144, 149. Rimma Gerlovina and Valeriy Gerlovin: 48–49a. Galerie Got, Paris: 107, 128. Giraudon, Vanves: 64. Hermann, Paris: 78. Hessische Landesbibliothek, Darmstadt, Germany: 52. INRP Musée National de l'Education, Rouen, France: 60, 61, 62–63, 63, 153. Jacana, Paris: 23. Kunsthistorisches Museum, Vienna: 44, 45. Kunstmuseum Basel: 66a–67a. Lauros-Giraudon, Vanves, France: 62, 76b. Lotos Film Thiem, Kaufbeuren: 24. Métis, Paris, Pascal Dolémieux: 70a. Metropolitan Museum of Art, New York: 65. Roland and Sabrina Michaud, Paris: 80b, 110–11. MNAM, Paris: 14–15, 107, 124. Pace-Wildenstein: 99b. Pedicini, Naples: 84b. Eric Pollitzer, Garden City Park, N.Y.: 133. Private collection, Paris: 92. Rapho, Paris, Roland Michaud: 26–27/Paolo Koch 56b. Rashed, Paris 73, 88. RMN, Paris: 16, 28–29, 32b, 33, 57. Roger-Viollet, Paris: 99a, 100–101, 138, 155. Scala, Florence: 108b. © Sevenarts Ltd., London: 68, 69, 106. Siné, Paris: 123. Roland Topor, Paris: 74–75. Reprinted by permission of the *Wall Street Journal*, © 1997 Dow Jones & Company, Inc. All Rights Reserved Worldwide: 126–27.

Text Credits

Edwin A. Abbott, *Flatland: A Romance of Many Dimensions,* New York, Dover Publications, Inc., 1992, reprinted by permission. Aristotle, *Metaphysics,* trans. © Hippocrates G. Apostle, 1966. Carl B. Boyer, *A History of Mathematics,* 2d edition (with Uta C. Merzbach), Copyright © 1968, 1989, 1991 by John Wiley & Sons, Inc., Reprinted by permission. Scott Buchanan, *Poetry and Mathematics,* Chicago, The University of Chicago Press, 1975, © Scott Buchanan, 1929, 1957, 1962. Boris A. Kordemsky, *The Moscow Puzzles,* ed. Martin Gardner, trans. Albert Parry, New York, Dover Publications, Inc., 1971, 1972, reprinted by permission. Lao Tzu, *Tao Te Ching,* trans. H. G. Ostwald, London, Arkana (Penguin Books), 1989. Dylan Loeb McClain, "The Evolution of the Calculator," Copyright © 1997 by The New York Times Company, Reprinted by permission. Richard Phillips, *Numbers: Facts, Figures and Fiction,* New York, N.Y., Cambridge University Press, 1994. Philolaus, in *The Presocratics,* trans. Philip Wheelwright, Bobbs-Merrill Co., 1966. Jean Piaget, *Psychology and Epistemology,* trans. Arnold Rosin, copyright © by Editions Denoel, 1970. Reprinted by permission of Georges Borchardt, Inc., U.S. and Canada, 1971. Plato, *Republic,* trans. Paul Shorey, in *The Collected Dialogues of Plato, Including the Letters,* ed. Edith Hamilton and Huntington Cairns, © 1989 Princeton University Press. Henri Poincaré, *Science and Hypothesis,* New York, Dover Publications, Inc., 1952, reprinted by permission. Jean-Philippe Rameau, *Treatise on Harmony,* trans. Philip Gossett, New York, Dover Publications, Inc., 1971, reprinted by permission. Edward Rothstein, *Emblems of Mind: The Inner Life of Music and Mathematics,* New York, Times Books, Random House, 1995. Bertrand Russell, *The Principles of Mathematics,* second edition. © 1938 by Bertrand Russell, Reprinted by permission of W. W. Norton & Company, Inc., New York, and Routledge, London, 1996. *The Song of God: Bhagavad-Gita,* trans. Swami Prabhavananda and Christopher Isherwood, New York, Mentor (Penguin Books), 1944, 1951, 1972, Reprinted by permission of the Vedanta Society of Southern California. Paul Valéry, *Masters and Friends,* trans. Martin Turnell, Princeton University Press, 1968.

Denis Guedj is professor of the history of science at
the University of Paris–VIII, where he has also
taught mathematics and film, as well as a filmmaker
and performer. He is author of numerous books on
the history of mathematics and science.

Translated from the French by Lory Frankel

Mathematics consultant: Joseph Wickham

First published in the United Kingdom in 1998
by Thames & Hudson Ltd, 181A High Holborn,
London WC1V 7QX

Reprinted in 2002

British Library Cataloguing-in-Publication Data

A catalogue record for this book
is available from the British Library

ISBN 0-500-30080-1

Printed and bound in Italy
by Editoriale Lloyd, Trieste